SpringerBriefs in Physics

W0112274

For further volumes:
http://www.springer.com/series/8902

Francesco Catoni · Dino Boccaletti ·
Roberto Cannata · Vincenzo Catoni ·
Paolo Zampetti

Geometry of Minkowski Space–Time

 Springer

Francesco Catoni
Via Veglia 10
00141 Rome
Italy
e-mail: paolo.zampetti@enea.it

Vincenzo Catoni
Via Veglia 10
00141 Rome
Italy
e-mail: vjncenzo@yahoo.it

Dino Boccaletti
Department of Mathematics
University of Rome "La Sapienza"
Piazzale Aldo Moro 2
00185 Rome
Italy
e-mail: Boccaletti@uniroma1.it

Paolo Zampetti
Casaccia Research Centre
ENEA
Via Anguillarese 301
00123 Rome
Italy
e-mail: paolo.zampetti@enea.it

Roberto Cannata
Casaccia Research Centre
ENEA
Via Anguillarese 301
00123 Rome
Italy
e-mail: roberto.cannata@enea.it

ISSN 2191-5423

e-ISSN 2191-5431

ISBN 978-3-642-17976-1

e-ISBN 978-3-642-17977-8

DOI 10.1007/978-3-642-17977-8

Springer Heidelberg Dordrecht London New York

Cover design: eStudio Calamar, Berlin/Figueres

Printed on acid-free paper

Springer is part of Springer Science+Business Media (www.springer.com)

Preface

We have written this book with the intention of providing the students (and the teachers) of the first years of university courses with a tool which is easy to be applied and allows the solution of any problem of relativistic kinematics at the same time.

The novelty of our presentation consists of the extensive use of hyperbolic numbers for a complete formalization of the kinematics in the Minkowski space–time.

In other words, in this book the mathematical relation, stated by special relativity, between space and time is formalized.

We recall from Paul Davies book [1], the different significances attributed to "time" over the centuries:

For millennia the traditional cultures have given to time an intuitive meaning. Its cyclic nature and biological rhythms predominate over its measure and time and eternity are complementary concepts.

Before Galileo and Newton, the time was subjective, not a parameter we have to measure with geometrical precision.

Newton encapsulated it in the World description just as a parameter for the mathematical description of the motion: practically the time did nothing.

Einstein has given it again its place in the heart of the Nature, as a fundamental part of the physics.

Einstein did not complete this revolution that, unfortunately, remained unfinished.

To the last sentence Einstein would have most probably replied [2] that the physical laws will never be the definitive ones. All scientists can step forwards in the advancement of the scientific knowledge, but they are sure that the results obtained cannot be the definitive ones.

The achievement of special relativity, to which P. Davies refers, can be summarized as: *space and time must be considered as equivalent quantities for the description of physical laws.*

In this book we formalize this equivalence.

Actually, even if just after the formulation of special relativity, Hermann Minkowski proposed (1907–1908) that the relation between space and time can be considered as a new geometry, but a mathematics that would allow us to operate with this geometry as we do with Euclidean geometry was not formalized yet.

This formalization has been carried out by the authors in a series of papers [3, 4], later rearranged in a book [5] where, besides the well-established aforesaid formalization, some themes of research are proposed.

The aim of the present book is supplying the tools for solving problems in space–time in the same "automatic way" as problems of analytic geometry and trigonometry are solved in secondary schools. The previous knowledge of mathematics which is required is the same required in the first year of scientific University courses.

Further we show the basic ideas of our treatment and how these ideas derive from the "scientific revolutions" of 19th and 20th centuries: in particular, the necessary link between mathematics and physics and the synergic effects that allow their development.

The papers and the books listed in the bibliography are not indispensable for understanding the contents of the book, but they can help people who want to carry on further research. Actually, even if the mathematics used can be considered elementary, the topics dealt with are the subject of research in progress. What we are saying is that an appropriate reconsideration of some points of elementary mathematics and geometry can be the starting point for obtaining original and valuable results.

Rome, January 2011

Francesco Catoni
Dino Boccaletti
Roberto Cannata
Vincenzo Catoni
Paolo Zampetti

References

1. P. Davies, About time, Orion Production (1995)
2. A. Einstein, L. Infeld, *The Evolution of Physics* (Simon and Shuster, New York, 1966)
3. F. Catoni, R. Cannata, V. Catoni, P. Zampetti, Hyperbolic trigonometry in two-dimensional space-time geometry. Nuovo Cimento. B. **118**(5), 475 (2003)
4. D. Boccaletti, F. Catoni, V. Catoni, The Twins Paradox for uniform and accelerated motions. Adv. Appl. Clifford Al. **17**(1), 1 (2007); The Twins Paradox for non-uniformly accelerated motions. Adv. Appl. Clifford Al. **17**(4), 611 (2007)
5. F. Catoni, D. Boccaletti, R. Cannata, V. Catoni, E. Nichelatti, P. Zampetti, *The Mathematics of Minkowski Space-Time* (Birkhäuser Verlag, Basel, 2008)

Contents

Chapter 1
Introduction

It is largely known that the Theory of Special Relativity was born as a consequence of the demonstrated impossibility for the Maxwell's electromagnetic (e.m.) theory of obeying Galilean transformations. The non-invariance of the e.m. theory under Galilean transformations induced the theoretical physicists, at the end of the twelfth century, to invent new space–time transformations which did not allow to consider the time variable as "absolutely" independent of the space coordinates. It was thus that the transformations which, for the sake of brevity, today we call Lorentz transformations were born.[1]

A consequence of the choice of the Lorentz transformations as the ones which keep the e. m. theory invariant when passing from an inertial system to another one (e.g. the e.m. waves remain e.m. waves) was to revolutionize the kinematics, i.e. the basis from which one must start for building the mechanics. It was no longer possible, at least for high velocities, to use Galilean transformations: when passing from an inertial system to another one, also the time variable (until when considered as "the independent variable") was forced to be transformed together with the space coordinates.

After the works of Lorentz and Poincaré, it was Einstein's turn to build a relativistic mechanics starting from the kinematics based on Lorentz transformations.

However, we owe to Hermann Minkowski the creation of a four-dimensional geometry in which the time entered as the fourth coordinate: the Minkowski space–time as it is called today.

We recall that for a long period of time the introduction of "i t" (where "i" is the imaginary unit) as the fourth coordinate was into use with the aim of providing to space–time a pseudo-Euclidean structure.

We shall see in this book that it is not the introduction of an imaginary time, but of a system of numbers (the hyperbolic numbers) related in many respects with complex numbers, that can describe the relation (symmetry) between space and

---

[1] An exhaustive account of the subject and its historical context can be found in the book—Arthur I. Miller: Albert Einstein's Special Theory of Relativity—Springer, 1998

F. Catoni et al., *Geometry of Minkowski Space–Time*, SpringerBriefs in Physics,
DOI: 10.1007/978-3-642-17977-8_1, © Francesco Catoni 2011

time. Moreover this system of numbers allows one a mathematical formalization that, from a logical point of view (an axiomatic–deductive method starting from axioms of empirical evidence) as well as from a practical one (the problems are solved in the same automatic way as the problems of analytical geometry and trigonometry are), is equivalent to the analytical formalization of Euclidean geometry.

Even though Minkowski already introduced hyperbolas in place of calibration circles and a copious literature on the subject does exist, until now a formalization, rigorous and extremely simple at the same time, was not obtained.

In this book we explain how to treat any problem of relativistic kinematics. The expounded formalization allows one to reach an exhaustive and non-ambiguous solution.

The final appendix, while contains a short outline on the evolution of the concept of geometry, helps us to reflect upon the nature of the operation made by the authors: a Euclidean way for facing a non-Euclidean geometry.

Chapter 2
Hyperbolic Numbers

Abstract Complex numbers can be considered as a two components quantity, as the plane vectors. Following Gauss complex numbers are also used for representing vectors in Euclidean plane. As a difference with vectors the multiplication of two complex numbers is yet a complex number. By means of this property complex numbers can be generalized and hyperbolic numbers that have properties corresponding to Lorentz group of two-dimensional Special Relativity are introduced.

Keywords Complex numbers · Gauss-Argand · Generalization of complex numbers · Hyperbolic numbers · Space-time geometry · Lorentz group

Complex numbers represent one of the most intriguing and emblematic discoveries in the history of science. Even if they were introduced for an important but restricted mathematical purpose, they came into prominence in many branches of mathematics and applied sciences. This association with applied sciences generated a synergistic effect: applied sciences gave relevance to complex numbers and complex numbers allowed formalizing practical problems. A similar effect can be found today in the "system of hyperbolic numbers", which has acquired the meaning and importance as the *Mathematics of Special Relativity*, as shown in this book.

Let us recall some points from the history of complex numbers and their generalization.

Complex numbers are today introduced with the purpose of extending the field of real numbers and for having always two solutions for the second degree equations and, as an important applicative example, we recall the Gauss *Fundamental theorem of algebra* stating that "all the algebraic equations of degree N has N real or imaginary roots". Further Gauss has shown that complex and real numbers are adequate for obtaining all the solutions for any degree equation.

Coming back to complex numbers we now recall how their introduction has a practical reason. Actually they were introduced in the 16th century for solving a mathematical paradox: to give a sense to the real solutions of cubic equations that appear as the sum of square roots of negative quantities (see Sect. 2.6.1). Really the goal of mathematical equations was to solve practical problems, in particular geometrical problems, and if the solutions were square roots of negative quantities, as can happen for the second degree equations, it simply meant that the problem does not have solutions. Therefore it was unexplainable that the real solutions of a problem were given by some "imaginary quantities" as the square roots of negative numbers.

Their introduction was thorny and the square roots of negative quantities are still called *imaginary numbers* and contain the symbol "i" which satisfies the relation $i^2 = -1$. *Complex numbers* are those given by the symbolic sum of one real and one imaginary number $z = x + iy$. This sum is a symbolic one because it does not represent the usual sum of "homogeneous quantities", rather a "two components quantity" written as $z = \mathbf{1}x + iy$, where $\mathbf{1}$ and i identify the two components.

Today we know another two-component quantity: the plane vector, which we write $\mathbf{v} = \mathbf{i}x + \mathbf{j}y$, where $\mathbf{i}$ and $\mathbf{j}$ represent two unit vectors indicating the coordinate axes in a Cartesian representation. Despite there being no a priori indication that a complex number could represent a vector on a Cartesian plane, complex numbers were the first representation of two-component quantities on a Cartesian (or Gauss–Argand) plane (see Fig. 2.1), and they are also used for representing vectors in a Euclidean Cartesian plane.

Now we can ask: what are the reasons that allow complex numbers to represent plane vectors? The answer to this question has allowed us to formalizing the geometry and trigonometry of Special Relativity space–time.

2.1 The Geometry Associated with Complex Numbers

Let us now recall the properties that allow us to use complex numbers for representing plane vectors. The first property derives from the invariant of complex numbers the modulus, indicated with $|z|$, and given by $|z| = \sqrt{(x + iy)(x - iy)} \equiv \sqrt{x^2 + y^2}$.

An important property of the modulus is: given two complex numbers z_1, z_2, we have $|z_1 \cdot z_2| = |z_1| \cdot |z_2|$.

If we represent the complex number $x + iy$ as a point $P \equiv (x, y)$ of the Gauss–Argand plane (Fig. 2.1), the quantity $\sqrt{x^2 + y^2}$ represents the distance of P from the coordinates origin. This quantity is invariant with respect to translations and rotations of the coordinate axes. Now if in $z = x + iy$, we give to 1 and i the same meaning of $\mathbf{i}, \mathbf{j}$ in the vectors representation, $|z|$ is the modulus of the vector.

In addition another relevant property allows complex numbers representing plane vectors and the related linear algebra. Actually let us consider the product of a complex constant, $a = a_r + ia_i$ by a complex number:

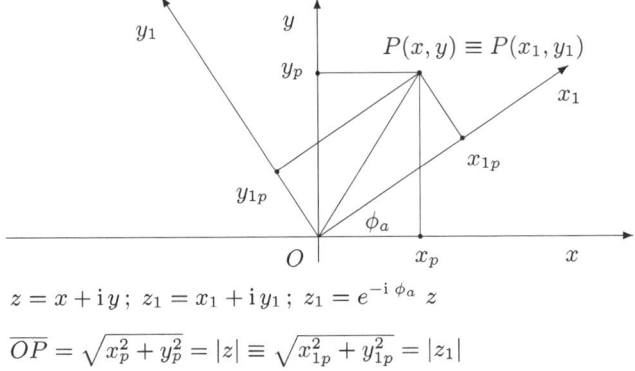

$$z = x + \mathrm{i}\, y\,;\; z_1 = x_1 + \mathrm{i}\, y_1\,;\; z_1 = e^{-\mathrm{i}\,\phi_a}\, z$$

$$\overline{OP} = \sqrt{x_p^2 + y_p^2} = |z| \equiv \sqrt{x_{1p}^2 + y_{1p}^2} = |z_1|$$

Fig. 2.1 Gauss representation of complex numbers. The square roots of negative numbers are called "imaginary" and are preceded by the symbol "i", which satisfies the relation $\mathrm{i}^2 = -1$. The expressions $z = x + \mathrm{i}y$, given by the symbolic sum of a real and an imaginary number are called *complex numbers*. We call this sum "symbolic" since it does not represent the usual sum between homogeneous quantities, rather it is a "two component quantity", written as: $z = \mathbf{1}x + \mathrm{i}y$, where $\mathbf{1}$ and $\mathbf{i}$ identify the two components. Gauss represented these numbers on a Cartesian plane x, y, associating with the complex number the point P with abscissa x and ordinate y. This "strange representation" can derive from the fact that probably Gauss noted that the product between $z = x + \mathrm{i}y$ and the particular number, called the *complex conjugate* $\bar{z} = x - \mathrm{i}y$ is the real number given by $z \cdot \bar{z} = (x + \mathrm{i}y)(x - \mathrm{i}y) = x^2 + y^2$, that also represent the Euclidean distance of P from the coordinate origin. The square root of this quantity, written as $|z|$, is called the modulus and is characteristic of the complex number. The modulus satisfies the relation: given two complex numbers z_1, z_2, we have $|z_1 \cdot z_2| = |z_1| \cdot |z_2|$, from which another link with Euclidean geometry follows. Actually let us consider the complex constant $a = \cos \phi_a + \mathrm{i}\sin \phi_a$, that can be written (2.4) as $a = e^{\mathrm{i}\phi_a}$, therefore $|a| = e^{\mathrm{i}\phi_a} \cdot e^{-\mathrm{i}\phi_a} = 1$. By considering the product $z_1 = az$, we have $|z_1| = |a||z| = |z|$. This transformation preserves the modulus. Developing the transformation we have: $x_1 + \mathrm{i}y_1 = (\cos \phi_a + \mathrm{i}\sin \phi_a)(x + \mathrm{i}y) = x\cos \phi_a - y\sin \phi_a + \mathrm{i}(x\sin \phi_a + y\cos \phi_a)$. Making equal the real and the imaginary terms, we obtain the relations between the coordinates of the point P after the rotation of the segment $\overline{OP}$ around the axes origin O of an angle ϕ_a. These same expressions represent the transformation of the coordinates of a point in the Cartesian plane, when the reference axes are rotated by the angle $-\phi_a$. We also have $\mathrm{i} = \exp[\mathrm{i}\pi/2]$, then the axes x and y are, automatically orthogonal. These properties show an "unimaginable" correspondence between complex numbers and Euclidean geometry

$$z_1 \equiv x_1 + \mathrm{i}y_1 = az \equiv (a_r + \mathrm{i}a_i)(x + \mathrm{i}y). \tag{2.1}$$

By considering another constant $b = b_r + \mathrm{i}b_i$, we have $z_2 = bz_1 \equiv baz \equiv cz$. Since c, i.e., the result of product between a and b is yet a complex constant, the product of z for a complex constant is a group (see Appendix A.4), called *multiplicative group* [1, Chap. 3]. Now we note that (2.1), is equivalent to the expression of linear algebra

$$\begin{pmatrix} x_1 \\ y_1 \end{pmatrix} = \begin{pmatrix} a_r & -a_i \\ a_i & a_r \end{pmatrix} \begin{pmatrix} x \\ y \end{pmatrix}. \tag{2.2}$$

In particular, the complex number plays the role of both a vector and an operator (matrix). Actually, the constant a is written in matrix form (like the operators in linear algebra), while z is represented as a column vector.

Now we look for the geometrical meaning of this multiplicative group, beginning with a different representation of complex numbers that starts from the famous Euler formula

$$\exp[i\phi] = \cos \phi + i \sin \phi. \tag{2.3}$$

This formula is very important for the following of the book and for this reason we think advisable to recall its first demonstration given by Euler. Actually Euler applied to a complex quantity the series development that is true for the real exponential function and realizes that the real and imaginary terms correspond to the series expansion of cosine and sine functions:

$$\exp[i\phi] = \sum_{l=0}^{\infty} \frac{(i\phi)^l}{l!} = \sum_{l=0}^{\infty} \frac{(i\phi)^{2l}}{(2l)!} + \sum_{l=0}^{\infty} \frac{(i\phi)^{2l+1}}{(2l+1)!}$$

$$= \sum_{l=0}^{\infty} (-1)^l \frac{(\phi)^{2l}}{(2l)!} + i \sum_{l=0}^{\infty} (-1)^l \frac{(\phi)^{2l+1}}{(2l+1)!} = \cos \phi + i \sin \phi. \tag{2.4}$$

Actually, in Euler's time the theory of power series was not sufficiently developed. Therefore it was not known that the displacement of terms, necessary for bringing together the real and imaginary terms, is possible only for absolutely convergent series, a property that the series (2.4) holds. Therefore his procedure is today considered mathematically correct. From (2.4) it follows

$$\exp[-i\phi] = \cos \phi - i \sin \phi. \tag{2.5}$$

By multiplying (2.3) · (2.5) we obtain, in an algebraic way, the well known trigonometric relation

$$1 \equiv \exp[i\phi] \cdot \exp[-i\phi] \equiv (\cos \phi + i \sin \phi)(\cos \phi - i \sin \phi)$$
$$= \cos^2 \phi + \sin^2 \phi. \tag{2.6}$$

Summing and subtracting (2.4) and (2.5) we obtain a formal relation between the trigonometric functions and the exponential of an imaginary quantity

$$\cos \phi = \frac{\exp[i\phi] + \exp[-i\phi]}{2}; \quad \sin \phi = \frac{\exp[i\phi] - \exp[-i\phi]}{2i}. \tag{2.7}$$

We have called (2.7) a formal relations since we cannot give a meaning to the exponential of an imaginary quantity. In any case we see in the following its extension and its relevance.

The introduction of the exponential function of imaginary quantities allows us to introduce the exponential transformation

$$x + iy = \exp[\rho' + i\phi] \equiv \exp[\rho'](\cos\phi + i\sin\phi), \tag{2.8}$$

and setting $\exp[\rho'] = \rho$ we obtain the *polar transformation*

$$x + iy = \rho \exp[i\phi] \equiv \rho(\cos\phi + i\sin\phi). \tag{2.9}$$

$\rho = \sqrt{x^2 + y^2} = |z|$ is called *radial coordinate*, and $\phi = \tan^{-1}[y/x]$ *angular coordinate*. If we write the constant a of (2.1) in polar form,

$$a \equiv (a_r + ia_i) = \rho_a(\cos\phi_a + i\sin\phi_a),$$

where $\rho_a = \sqrt{a_r^2 + a_i^2}$; $\phi_a = \tan^{-1}[a_i/a_r]$, (2.2) becomes

$$\begin{pmatrix} x_1 \\ y_1 \end{pmatrix} = \rho_a \begin{pmatrix} \cos\phi_a & -\sin\phi_a \\ \sin\phi_a & \cos\phi_a \end{pmatrix} \begin{pmatrix} x \\ y \end{pmatrix} \equiv \rho_a \begin{pmatrix} x\cos\phi_a - y\sin\phi_a \\ x\sin\phi_a + y\cos\phi_a \end{pmatrix}. \tag{2.10}$$

We see that the constant a plays the role of an operator representing a homogeneous dilatation ρ_a (homothety) and the transformation for the coordinates of a point P in a rotation, of an angle ϕ_a, around the coordinates origin. Or, changing $\phi_a \to -\phi_a$, for an orthogonal-axis rotation.

If $\rho_a = 1$, and if we add another constant $b = b_r + ib_i$, then $z_1 = az + b$ gives the permissible vector transformations in a Euclidean plane. For these transformations we have $|z_1| = |a||z| = |z|$, i.e., the modulus of complex numbers or vectors (or the length of a segment) is invariant.

Then, *the additive and unitary multiplicative groups of complex numbers are equivalent to the Euclidean groups of rotations and translations, which depends on the three parameters ϕ_a, b_r, b_i and, as shown in Fig. 2.1, complex numbers can be used for describing plane-vector algebra.* Now we can ask if other systems of numbers have similar properties. For inquiring into this possibility we begin by comparing the algebraic properties of sum and product for complex numbers with the ones of plane vectors:

As the sum is concerned it is defined in the same way, for both complex numbers and vectors, as the sum of the components, i.e., given the complex numbers $z_1 = x_1 + iy_1$ and $z_2 = x_2 + iy_2$, we have

$$z_1 + z_2 = (x_1 + x_2) + i(y_1 + y_2).$$

In the definition of the product there is a relevant difference:

1. the product between two complex numbers is the same as for real numbers just by adding the rule $i^2 = -1$: we have $z_1 z_2 \equiv (x_1 + iy_1)(x_2 + iy_2) = x_1 x_2 - y_1 y_2 + i(x_1 y_2 + x_2 y_1)$. Therefore *the product of two complex numbers is yet a complex number.*
2. As the plane vectors are concerned, their product is not a vector but rather it is a new quantities derived from physics. In particular the *scalar* and the *vector* products are defined.

So we can summarize: the product between vectors is a new quantity while the product between complex numbers is yet a complex number: *complex numbers are a group also with respect to the product operation.*

Now we recall how this property allows us to generalize the complex numbers [2]. This research for generalization can look as opposed to Gauss theorem that stated it is not necessary the introduction of new number systems more than real and complex numbers, but Gauss referred to solutions of algebraic equations that was the purpose of the introduction of complex numbers. Differently we are now looking for new uses of complex numbers, anyway we see that these systems are related with the kind of roots of the second degree equations.

2.2 Generalization of Complex Numbers

Let us consider a two components quantity written as the complex numbers $z_1 = x_1 + uy_1$ and $z_2 = x_2 + uy_2$, where u represents a general **versor**[1] for which we have not, a priori, defined the multiplication rule, i.e., the meaning of u^2 and, as a consequence, of all the powers of u. For the product we have

$$z_3 \equiv z_1 z_2 \equiv (x_1 + uy_1)(x_2 + uy_2) = x_1 x_2 + u(x_1 y_2 + x_2 y_1) + u^2 y_1 y_2, \quad (2.11)$$

we say that z is *a generalized complex numbers, if the result of this multiplication is a number of the same kind*

$$z_3 = q_1(x_1, x_2, y_1, y_2) + u q_2(x_1, x_2, y_1, y_2), \quad (2.12)$$

where q_1, q_2 are quadratic forms as function of the components.

We obtain this result setting u^2 as a linear combination of 1 and of the versor u:
$u^2 = \alpha + u\beta; \alpha, \beta \in \mathbf{R}$ [4]. Actually with this position (2.11) becomes

$$z_3 = x_1 x_2 + \alpha y_1 y_2 + u(x_1 y_2 + x_2 y_1 + \beta y_1 y_2) \quad (2.13)$$

In this way the generalized complex numbers are a group with respect to the product. These numbers are also indicated by

$$\{z = x + uy; \ u^2 = \alpha + u\beta; \ x, y, \alpha, \beta \in \mathbf{R}, \ u \notin \mathbf{R}\}, \quad (2.14)$$

---

[1] The name *versor* has been firstly introduced by Hamilton for the unitary vectors of his quaternions [3]. This name derives from the property of the imaginary unity "i" since, as can be seen from Euler formulas, multiplying by "i" is equivalent to "rotate", in a Cartesian representation, the complex number of $\pi/2$. Since this property also holds for the hyperbolic numbers that we are going to introduce, we use this name that also states the difference with the unitary vectors of linear algebra.

In the theory of hypercomplex numbers [1, Chap. 2], the constants α, β are called **structure constants** and, as we see in this two-dimensional example, *from their values derive the properties of the three systems of two-dimensional numbers.*

It is known that complex numbers are considered as an extension of real numbers, as regarding the division. Actually as for real numbers it is ever possible except for the null element $x = 0, y = 0$. Now we see how the general complex numbers can be classified by means of their property about the division.

Actually for the division, given a number $a + ub$, one has to look for a number $z = x + uy$ such that

$$(a + ub)(x + uy) = 1. \tag{2.15}$$

If (2.15) is satisfied, the inverse z of $a + ub$ exists and we can divide any number by $a + ub$, by multiplying it by z. Thanks to the multiplication rule (2.14), (2.15) is equivalent to the real system obtained by equating the coefficients of the versors 1 and u

$$\begin{aligned} ax + \alpha by &= 1, \\ bx + (a + \beta b)y &= 0, \end{aligned} \tag{2.16}$$

As it is known from the theory of linear systems, (2.16) has a solution if the determinant of the coefficients, given by

$$D = a^2 + \beta ab - \alpha b^2 \equiv \left(a + \frac{\beta}{2}b\right)^2 - \left(\alpha + \frac{\beta^2}{4}\right)b^2, \tag{2.17}$$

is different from zero. Actually if $D = 0$ the associated homogeneous system $ax + \alpha by = 0, bx + (a + \beta b)y = 0$ admits non-null solutions. These numbers for which the product between two non-null numbers $a + ub$ and $x + uy$, is zero are called **divisors of zero** [1, Chap. 2]. The origin of this name derives from the exposed considerations: actually, for these numbers we can formally write

$$a + ub = \frac{0}{x + uy},$$

therefore dividing zero for $x + uy$ we obtain the finite quantity $a + ub$.

For studying this property (2.17), in the last passage, has been divided into two terms: now we inquire into the second one, by setting

$$\Delta = \beta^2 + 4\alpha \tag{2.18}$$

we see that the sign of the real quantity Δ determines the possibility of executing the division between two numbers. Actually, let us consider in the (β, α) plane the parabola, obtained by setting $\Delta = 0$

$$\alpha = -\beta^2/4, \tag{2.19}$$

that divides the plane into three regions (see Fig. 2.2). In these regions we have $\Delta > 0, \Delta = 0, \Delta < 0$.

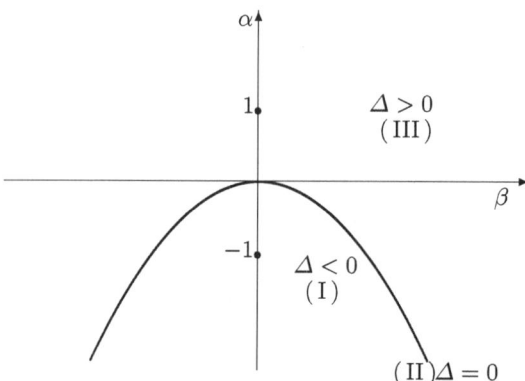

Fig. 2.2 The three types of two-dimensional algebras. Complex numbers can be considered as a two components quantity and are used to representing plane vectors. On the other hand, together with this correspondence (Sect. 2.1) there is a relevant difference between complex numbers and vectors: the definition of the product. Actually for complex numbers, just adding the rule $i^2 = -1$, it is an extension of the product between real numbers, i.e., the result is a complex number. As the vectors are concerned, their product is not a vector but a new quantity introduced from physics. In particular the *scalar product* and the *vector product* are defined. *With respect to multiplication complex numbers are a group, vectors are not.* Thanks to this property complex numbers can be generalized and two other two-dimensional systems of numbers are introduced [4, 5]. Actually let us consider a two-components quantity we write as the complex numbers: $z = x + uy$, where u represents a generic versor. If we request that the product between two numbers z_1 and z_2 is a number of the same kind, i.e., z is a multiplicative group, u^2 must be defined as a linear combination of the versors 1 and u of the number z, i.e., we must set $u^2 = \alpha + u\beta$; $\alpha, \beta \in \mathbf{R}$ (Sect. 2.2). The values of α, β determine the properties of the system of numbers. In particular, by considering in the (β, α) plane the parabola $\alpha = -\beta^2/4$: the position of the point $P \equiv (\beta, \alpha)$ with respect to parabola, determines if the division, except for $x = y = 0$, is possible. This property allows us to classify the two-dimensional numbers into three types: (I) Inside the parabola ($\Delta < 0$) we call these systems **elliptic numbers**. In particular for $\beta = 0, \alpha = -1$ we have the **complex numbers**. The division is ever possible. (II) On the parabola ($\Delta = 0$), we call these systems **parabolic numbers**. The division is not possible for the numbers $2x + \beta y = 0$. (III) Outside the parabola ($\Delta > 0$), we call these systems **hyperbolic numbers**. The division is not possible for the numbers $2x + (\beta \pm 2\sqrt{\Delta})y = 0$. All these three systems have a geometrical or physical relevance. Actually complex numbers can represent Euclidean geometry, the group of parabolic numbers is equivalent to Galileo's group of classical dynamics [6] and hyperbolic numbers, as we at length see in this book, represent the Lorentz's group of Special Relativity

The position of a point $P \equiv (\beta, \alpha)$ with respect to parabola, determines three types of systems and we have

Theorem 2.1 *We can classify the general two-dimensional numbers into three classes according to the position of point* $P \equiv (\beta, \alpha)$ *with respect to parabola* (2.19).
The numbers of the same type have also in common:

1. *the characteristic property of the modulus, i.e., the definition of distance that relates the system of numbers with a geometry;*
2. *the topological properties of the representative plane.*

Proof By referring to Fig. 2.2, we have

1. If $P \in (I), \Delta < 0$ and D, as the sum of two squares, is never negative and it is equal to 0 just for $a = b = 0$. Therefore any non-null element has an inverse and, as a consequence, division is possible for any non-null number. These systems are called **elliptic numbers**.

2. If P is on the parabola, $D = (a + \beta b/2)^2$ is zero if a, b are on the straight line $a + \beta b/2 = 0$. Each of them admits divisors of zero satisfying $x + (\beta/2)y = 0$. Division is possible for all the other numbers. These systems are called **parabolic numbers**.

3. If $P \in (III)$, the system (2.16) has solutions for a, b on the straight lines $a + (\beta \pm \sqrt{\Delta})b/2 = 0$. Each of them admits divisors of zero satisfying $x + (\beta \pm \sqrt{\Delta})/2y = 0$. Division is possible for all the other numbers. These systems are called **hyperbolic numbers**.

   The divisor of zero determines the topology of the representative plane: this plane is divided in four sectors.   □

We can say that *the types of the general two-dimensional systems derive from the kinds of solutions of the second degree equation* (2.17) *in* a/b, *obtained by setting* $D = 0$.

Let us now see the derivation of the nouns for the systems. Actually, let us consider the conic with equation

$$x^2 + \beta xy - \alpha y^2 = 0, \tag{2.20}$$

obtained from (2.17), by considering a, b as variables in Cartesian plane. According to whether $\Delta \equiv \beta^2 + 4\alpha$ is $< 0, = 0, > 0$, the curve is an ellipse, a parabola or a hyperbola. This is the reason for the names used for the three types of the general two-dimensional numbers. For the three cases, we define the **canonical systems** by setting $\beta = 0$ and

1. $\alpha \equiv u^2 = -1$. This is the case of the ordinary complex numbers. For these numbers we set, as usual $u \Rightarrow i$.

2. $\alpha \equiv u^2 = 0$.

3. $\alpha \equiv u^2 = 1$. This system is related to the pseudo-Euclidean (space–time) geometry, as we see in this book. For this system we set $u \Rightarrow h$. In this case the divisors of zero satisfy $y = \pm x$. In a Cartesian representation they are represented by the axes bisectors.

One can verify that any system can be obtained, from its canonical system, by a linear transformation of the versors (with the inverse transformation for the variables), i.e., it is isomorphic to the canonical system.

2.2.1 Definition of the Modulus

We note that the left-hand side of (2.20), for $\beta = 0, \alpha = -1$ represents the squared modulus of complex numbers, i.e., the real and invariant quantity obtained by multiplying $z \cdot \bar{z}$. Now we see how this quantity can be defined for a generic algebra. Actually looking at the last term of (2.17), we note that it can be written as $\Delta = 4\alpha + \beta^2 \equiv (2u - \beta)^2$: in this way D is the difference between two squared terms and can be written as the product of two linear terms that by substituting $a, b \Rightarrow x, y$ becomes $(x + uy)(x + \beta y - uy) \in \Re$. Therefore *for the generic algebra we define the number $\bar{z} = x + (\beta - u)y$, that multiplied for $x + uy$ gives a real quantity, as the complex conjugate of $z = x + uy$.*

We conclude this generalization of complex numbers by recalling the definition of [5, p. 11][2]: ◁ The fact that the most general complex numbers can be added, subtracted and multiplied, all the usual laws of these operations being conserved, but that it is not always possible divide one by another, is expressed by saying that such numbers form a ring. ▷

The representation of Euclidean geometry by means of complex numbers and the equivalence, from the algebraic point of view, between complex and hyperbolic numbers let us suppose that also the hyperbolic numbers, can be associated with a geometry. Now we see that their geometry is the one of special relativity.

2.3 Lorentz Transformations and Space–Time Geometry

We briefly recall how Lorentz transformations of Special Relativity were established.

For some decades, at the end of 19th century the Newton dynamics and gravitational theory together with Maxwell equations of electro-magnetic field were considered adequate for a complete description of physical world: the mechanics and gravitation law formalize the motions on the Earth and of celestial bodies, Maxwell equations, besides the technical and scientific relevance, also explain the light propagation.

Actually these two theories and the effort to put them in a same logical frame, brought about the starting ideas for the "scientific revolutions" of 20th century, that are today considered very far of being concluded.

We begin by setting out their different mathematical nature and the role the time holds.

1. The Newton dynamics equations give the bodies positions as functions of time. The time acts as a parameter.

---

[2] We use ◁...▷ to identify material that reports the original author's words or is a literal translation.

2. The Maxwell equations allow us to calculate the electric and magnetic fields, from static and moving charges. These fields depend in an equivalent way on space coordinates and time.

From a mathematical point of view the Newton equations are ordinary differential equations, Maxwell equations are a partial differential system.

Moreover besides these mathematical differences there were theoretical considerations and experimental results, as the Michelson and Morley experiment, that we directly recall, that stated that Newton and Maxwell equations are not equivalent also for a physical point of view.

The result of this debate was that Poincaré and Einstein, in the same year (1905), looked for the variable transformations that leaved the same expressions of Maxwell equations when one considers two reference systems in uniform relative motion. This requirement is the same as the invariance of Newton dynamics equations with respect to Galileo's group.

Both the scientists obtained the today known *Lorentz transformations* of special relativity.

Since Maxwell equations depend in an equivalent way from both time and space variables, also the transformations depend in an equivalent way on these variables.

For practical purposes the Lorentz transformations, notwithstanding can be considered as elementary from a mathematical point of view, have represented, for the connexion between space and time, a "revolution" with respect to settled philosophical concepts about "*time*".

The Poincaré and Einstein works reflect their professionalism and their interpretation of the results are complementary:

- Poincaré, one of the most important and encyclopedic mathematician at the turn of the century, associated these transformations with group theory (today known as Lorentz–Poincaré's group).
- Einstein, young physicist, had the cheek to extend the transformation laws relating space with time, obtained for Maxwell equations, to dynamics equations. This extension, together with the paper about photoelectric effect published by Einstein in the same year, was the basis for the most important scientific results of 20th century. Actually the results of both these works entail the equivalence between waves and corpuscles.

Now we briefly recall Einstein's formulation, who gives to the obtained transformations the physical meaning today accepted, in particular the extension to Newton dynamics of the obtained relation between space and time.

Einstein was able to obtain in a straightforward way and by means of elementary mathematics the today named Lorentz transformations, starting from the two postulates

1. all inertial reference frames must be equivalent
2. light's velocity is constant in all inertial systems.

The first postulate is the same stated by Galileo and applied to the laws of dynamics, that starts from the principle that by means of physical experiments we cannot detect the state of relative uniform motion.

The second one takes into account the results of experiments carried out by Michelson and Morley. These experiments have shown that *the speed of the light is the same in all inertial systems* and is independent of the direction of the motion of the reference frame relative to the ray of light. Actually also this experimental result, in contrast with the traditional physics, stimulated the search for new theories which could explain it.

For the formalization of the problem let us consider a reference system (t, x) and another (t_1, x_1) in motion with constant speed v_1 with respect to the first one. In this description t represents the time multiplied by light's velocity ($c = 1$) and v_1 the speed divided by c. The obtained transformations are

$$x_1 = \frac{x + v_1 t}{\sqrt{1 - v_1^2}}, \quad t_1 = \frac{v_1 x + t}{\sqrt{1 - v_1^2}}. \tag{2.21}$$

From these equations we note two relevant differences with respect to classical dynamics

1. both the length x and the time t depend on the speed of the reference frame in which are measured,
2. the square root of the quantity $t^2 - x^2 = t_1^2 - x_1^2$ is invariant and is called **proper time**.

The dependence of time on the speed of the reference system originated the "twin paradox". This problem is exhaustively formalized in Chap. 6.

Now let us consider a third system (t_2, x_2) in motion with speed v_2, with respect to system (t_1, x_1). The transformation equations from the first system and this one are again (2.21) with the substitution $v_1 \rightarrow v_T$ where v_T is given by

$$v_T = \frac{v_1 \pm v_2}{1 \pm v_1 v_2}, \tag{2.22}$$

where the $+$ and $-$ signs refer if v_1 and v_2 have the same or different directions. Therefore, as recalled in Appendix A.4, the relations (2.21) represent a group, i.e., two repeated transformations have the same expression as (2.21) with a speed v_T that is a function of the speeds (parameters) of the component transformations. From (2.22) it follows that if v_1 or v_2 are equal to 1 (one reference system have the speed of the light), we have $v_T = 1$. *The light's velocity is a limiting speed.*

Therefore a line in the t, x plane can represent the motion of a body only if the tangent lines have an angular coefficient $dx/dt \equiv v < 1$. These curves (or lines) are called **time-like**. In a similar way the lines that have an angular coefficient $dx/dt \equiv v > 1$ are called **space-like** and if $dx/dt \equiv v = 1$ are called **light-like**.

From the considerations of Appendix A it follows:

we can introduce a geometry that, with respect to the distance of Euclidean geometry, has as invariant the square root of the quantity $t^2 - x^2$, i.e., the proper time.

For this geometry the transformations (2.21) represent the motions.

This geometry is called pseudo-Euclidean or Minkowskian geometry.

2.4 The Geometry Associated with Hyperbolic Numbers

Let us now apply to hyperbolic numbers the same considerations that allow one to associate complex numbers with Euclidean geometry. Let us consider the system of hyperbolic numbers defined as

$$\{z = x + \mathrm{h}\,y; \mathrm{h}^2 = 1; x, y \in \mathbf{R}, \mathrm{h} \notin \mathbf{R}\}.$$

As for complex numbers, we call $\tilde{z} = x - \mathrm{h}\,y$ the hyperbolic conjugate[3] and define the modulus as $|z| = \sqrt{|z\tilde{z}|} \equiv \sqrt{|x^2 - y^2|}$. Therefore, by giving to one variable the physical meaning of *time*, the modulus can be recognized as the invariant of the two-dimensional special relativity (proper time). Now we see that *hyperbolic geometry is equivalent with the "geometry of special relativity"*. Let us now see its properties (Fig. 2.3).

As for complex numbers, given two hyperbolic numbers, z_1, z_2, we have $|z_1 \cdot z_2| = |z_1| \cdot |z_2|$.

Let us now consider the *multiplicative group* given by the product of z for a hyperbolic constant

$$z_1 = az \equiv (a_r + ha_h)(x + \mathrm{h}\,y), \tag{2.23}$$

this group can be expressed in vector–matrix form by

$$\begin{pmatrix} x_1 \\ y_1 \end{pmatrix} = \begin{pmatrix} a_r & a_h \\ a_h & a_r \end{pmatrix} \begin{pmatrix} x \\ y \end{pmatrix}. \tag{2.24}$$

This result is the same as (2.2): the multiplicative constant is an operator that acts on the vector (x, y).

As for complex numbers we can introduce the Hyperbolic exponential function and hyperbolic polar transformation.

In hyperbolic geometry these transformations play the same important role as the corresponding complex ones in Euclidean geometry. In [1, Chap. 7] the functions of a hyperbolic variable are introduced and analogies and differences with respect to functions of a complex variable are pointed out. Here we define the exponential function of a hyperbolic number following the method that Euler used for introducing the complex exponential with his famous formula (2.3). Actually going on as

---

[3] Here and in the following we use the symbol $\tilde{\ }$ for indicating the hyperbolic conjugate.

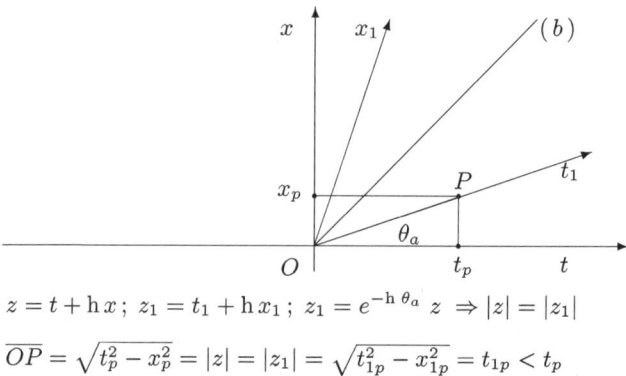

$$z = t + \mathrm{h}\,x \,;\; z_1 = t_1 + \mathrm{h}\,x_1 \,;\; z_1 = e^{-\mathrm{h}\,\theta_a}\,z \;\Rightarrow\; |z| = |z_1|$$

$$\overline{OP} = \sqrt{t_p^2 - x_p^2} = |z| = |z_1| = \sqrt{t_{1p}^2 - x_{1p}^2} = t_{1p} < t_p$$

Fig. 2.3 Geometry in pseudo-Euclidean plane The Euclidean geometry is defined by the invariance of geometrical figures with respect to their rotations and translations. Or, in a Cartesian representation, with respect to the reference axes rotations. These same two criterion can be applied to Lorentz transformations of special relativity. Following the Newton dynamics, the motion of a body is represented in a Cartesian reference frame x, y, by a curve expressed as function of a parameter t, to which we can give the physical meaning of *time*. From special relativity, formalized by (2.21), the "time" is "equivalent" to space then, in a representation on a plane, one reference axis must represent the time. Therefore, in this plane, a curve $x = x(t)$ represents the motion of a body. In particular a straight line $x = \beta t, \beta < 1$ represents a uniform motion. As we do for Euclidean geometry, represented in a Cartesian plane (Fig. 2.1), we can consider a second reference frame t_1, x_1 for which the relation between the old and new variables, corresponding to the Euclidean (2.10), is given by (2.21). In this transformation, as it has been shown for the first time by Minkowski after whom the space–time geometry is named, the transformed axes are not yet orthogonal. As it is shown in the figure, they go in a symmetric way toward the axes bisector (b) $t = x$. As we see in Fig. 3.2 these axes are yet "orthogonal in the hyperbolic geometry". Minkowski, by means of this geometrical representation, studied the dependence of t_1, x_1 on the body's velocity. The second possibility is the one studied in this book, i.e., to stay in a representative Cartesian plane by applying in this plane the geometry that leaves as invariant the "space–time" distance. We observe that while the axes rotation allow us to study just the uniform motions, represented by (2.21), this second approach allows us to quantify the "relativistic effect" for every motion. Actually for all the lines in t, x plane that represent a motion $(\beta < 1)$, its "relativistic length" is the *proper time*. These lengths can be evaluated for all the lines and then the differences or the ratios between the respective proper times. In particular (see Chap. 6) for uniform and uniformly accelerated motions, and all their compositions, these calculations are performed by elementary methods (Figs. 6.1–6.5); for a general motion by means of differential geometry (Fig. 6.6.)

for (2.4), taking into account that the even powers of h are equal to 1 and the odd powers to h, we can recognize that the real and hyperbolic parts of the series developments correspond to the hyperbolic trigonometric functions, and we have

$$
\begin{aligned}
\exp[\mathrm{h}\,\theta] &= \sum_{l=0}^{\infty} \frac{(\mathrm{h}\,\theta)^l}{l!} = \sum_{l=0}^{\infty} \frac{(\mathrm{h}\,\theta)^{2l}}{(2l)!} + \sum_{l=0}^{\infty} \frac{(\mathrm{h}\,\theta)^{2l+1}}{(2l+1)!} \\
&= \sum_{l=0}^{\infty} \frac{(\theta)^{2l}}{(2l)!} + \mathrm{h} \sum_{l=0}^{\infty} \frac{(\theta)^{2l+1}}{(2l+1)!} = \cosh\theta + \mathrm{h}\sinh\theta.
\end{aligned}
\tag{2.25}
$$

In particular $\cosh \theta = 1 +$ even powers, then $\cosh \theta > 1$.

By means of exponential function we introduce the exponential transformation

$$z \equiv x + h\, y = \exp[\rho' + h\, \theta] \equiv \exp[\rho'](\cosh \theta + h \sinh \theta), \qquad (2.26)$$

and setting $0 < \exp[\rho'] = \rho$ we obtain the *hyperbolic polar transformation*

$$z \equiv x + h\, y = \rho \exp[h\, \theta] \equiv \rho(\cosh \theta + h \sinh \theta). \qquad (2.27)$$

Where, as for complex numbers, ρ is called *radial coordinate*, and θ is called *angular coordinate*.

By comparing the real and the hyperbolic parts we can obtain, as for the polar transformation $\rho = \sqrt{x^2 - y^2} \equiv |z|$ and $\theta = \tanh^{-1}(y/x)$. From (2.27) we have

$$\tilde{z} \equiv x - h\, y = \rho \exp[-h\, \theta] \equiv \rho(\cosh \theta - h \sinh \theta). \qquad (2.28)$$

Setting $\rho = 1$ and multiplying (2.27) by (2.28) we obtain the relevant relations between the hyperbolic trigonometric functions

$$\begin{aligned} 1 &\equiv \exp[h\, \theta] \cdot \exp[-h\, \theta] \equiv (\cosh \theta + h \sinh \theta)(\cosh \theta - h \sinh \theta) \\ &= \cosh^2 \theta - \sinh^2 \theta. \end{aligned} \qquad (2.29)$$

From this relation and the previous one $(\cosh \theta > 1)$, it follows $\cosh \theta > \sinh \theta$, therefore the polar representation (2.27) holds for $x > y, x > 0$. In Chap. 3 we see how it is possible to extend it for representing points in the whole x, y plane.

Let us come back to the multiplicative groups (2.23) and, as for complex numbers, let us write the constant

$$a \equiv a_r + h\, a_h = \rho_a(\cosh \theta_a + h \sinh \theta_a) \qquad (2.30)$$

in hyperbolic polar form, (2.24) becomes

$$\begin{pmatrix} x_1 \\ y_1 \end{pmatrix} = \rho_a \begin{pmatrix} \cosh \theta_a & \sinh \theta_a \\ \sinh \theta_a & \cosh \theta_a \end{pmatrix} \begin{pmatrix} x \\ y \end{pmatrix} \equiv \rho_a \begin{pmatrix} x \cosh \theta_a + y \sinh \theta_a \\ x \sinh \theta_a + y \cosh \theta_a \end{pmatrix}. \qquad (2.31)$$

By considering, as for complex numbers, the transformation with $\rho_a = 1$ the modulus, i.e., the square root of the quantity $(x^2 - y^2)$, equivalent to proper time, is invariant. These transformations shall be called hyperbolic rotations and, we are going to see, they represent the Lorentz transformations of Special Relativity.

2.4.1 Hyperbolic Rotations as Lorentz Transformations

Let us write a space–time vector as a hyperbolic variable,[4]$w = t + \mathrm{h}\,x$ and consider a unitary hyperbolic constant $a = a_r + \mathrm{h}\,a_\mathrm{h}; a_r^2 - a_h^2 = 1$. If we give to the components of constant a the physical meaning given to the variables, a_r corresponds to time and a_h to a space variable. Therefore $a_\mathrm{h}/a_r \equiv x/t$ has the meaning of a velocity v. If a represents a physical motion $(v < 1)$, it must be $a_r > a_h$ (with this position a is a time-like constant). Setting a in polar form (2.30), we have

$$a_r + \mathrm{h}\,a_\mathrm{h} \equiv \exp[\mathrm{h}\,\theta_a] \equiv \cosh\theta_a + \mathrm{h}\sinh\theta_a$$
$$\text{where} \quad \theta_a = \tanh^{-1}[a_\mathrm{h}/a_r] \equiv \tanh^{-1}[v]. \tag{2.32}$$

Transformation (2.31) becomes

$$t_1 + \mathrm{h}\,x_1 = t\cosh\theta_a + x\sinh\theta_a + \mathrm{h}(t\sinh\theta_a + x\cosh\theta_a). \tag{2.33}$$

By considering as equal the coefficients of versors "1" and "h", as we do in complex analysis, we get the Lorentz transformation of two-dimensional special relativity [7]. Actually, from the second of (2.32) we have $\tanh\theta_a = v$, and, by means of the relation (2.29), we have

$$\sinh\theta_a = \frac{v}{\sqrt{1 - v^2}}, \quad \cosh\theta_a = \frac{1}{\sqrt{1 - v^2}}. \tag{2.34}$$

These relations allow us to verify that (2.33) are the same as (2.21).

In addition the composition (2.22) of speeds of two motions is given by the sum of the hyperbolic angles corresponding to the two speeds (see 4.24).

We also have

Theorem 2.2 *The Lorentz transformation is equivalent to a "hyperbolic rotation".*

Proof By writing the hyperbolic variable $t + \mathrm{h}\,x$ in exponential form (2.26)

$$t + \mathrm{h}\,x = \rho\exp[\mathrm{h}\,\theta],$$

the Lorentz transformation (2.33), becomes

$$t_1 + \mathrm{h}\,x_1 = a(t + \mathrm{h}\,x) \equiv \rho\exp[\mathrm{h}(\theta + \theta_a)]. \tag{2.35}$$

From this expression we see that the Lorentz transformation is equivalent to a "hyperbolic rotation" of the angle θ_a, of the $t + \mathrm{h}\,x$ variable. $\qquad\square$

---

[4] In all the problems which refer to Special Relativity (in particular in Chap. 6) we change the symbols by indicating the variables with letters reflecting their physical meaning $x, y \Rightarrow t, x$, i.e., t is a normalized time variable (light velocity $c = 1$) and x a space variable.

This correspondence allows us to call hyperbolic and pseudo-Euclidean the representative plane of space–time (Minkowski's) geometry and trigonometry. We note that to writing the Lorentz transformations by means of hyperbolic trigono- metric functions, is normally achieved by following a number of "formal" steps, i.e., by introducing an "imaginary" time $t' = it$ which makes the Lorentz invariant $(x^2 - t^2)$ equivalent to the Euclidean invariant $(x^2 + y^2)$, and by introducing the hyperbolic trigonometric functions through their equivalence with circular func- tions of an imaginary angle. We stress that this procedure is essentially formal, while the approach based on hyperbolic numbers leads to *a direct description of Lorentz transformation explainable as a result of symmetry (or invariants) pres- ervation*: the Lorentz invariant (space–time "distance") is the invariant of hyperbolic numbers and *the unimodular multiplicative group of hyperbolic num- bers represents the Lorentz transformations, as the unimodular multiplicative group of complex numbers represents the rotations in a Euclidean plane.*[5]

Therefore we conclude:

For the description of the physical world the hyperbolic numbers have the same relevance of complex numbers.

And, following Beltrami (see Sect. A.3), we can say: results that seem contradictory with respect to Euclidean geometry are compatible with another geometry as simple and relevant as the Euclidean one.

2.5 Conclusions

The association of hyperbolic numbers with the two-dimensional Lorentz's group of Special Relativity makes hyperbolic numbers relevant for physics and stimulate us to find their application in the same way as complex numbers are applied to Euclidean plane geometry.

In Chap. 4 we see that it is possible go over with respect to this project. Actually we show that the link between complex numbers and Euclidean geometry allow us to formalize, in a Cartesian plane, the trigonometric functions as a direct consequence of Euclid's rotation group (Sect. 4.1.1). This result allows us to show that all the trigonometry theorems can be obtained by an analytical method, as mathematical identities, instead of the usual method of Euclidean geometry and trigonometry for which theorems are demonstrated by means of the axiomatic- deductive method and geometrical observations. Afterward, taking into account that hyperbolic numbers have the same algebraic properties of complex numbers, these approaches to Euclidean geometry and trigonometry are extended to the space–time and, by means of hyperbolic numbers, the theorems are demonstrated

---

[5] Within the limits of our knowledge, the first description of Special Relativity, directly by these numbers was introduced by I. M. Yaglom [6].

through an algebraic method that replaces the absence, in space–time plane, of the intuitive Euclidean observations.

In this way, we obtain *the complete formalization of space–time geometry, by means of the axiomatic-deductive method, starting from experimental axioms, thus equivalent to Euclid's geometry construction.*

Therefore the problems in Minkowski space–time are solved as we usually do in the Euclidean Cartesian plane.

2.6 Appendix

2.6.1 Cubic Equation and Introduction of Complex Numbers

All the third degree equations were reduced to

$$x^3 + px + q = 0 \tag{2.36}$$

by the mathematicians Nicolò Fontana (named Tartaglia) and Girolamo Cardano (*Ars magna*, Nurberg, 1545) and they found the solution

$$x = \sqrt[3]{-\frac{q}{2} + \sqrt{\left(\frac{q}{2}\right)^2 + \left(\frac{p}{3}\right)^3}} + \sqrt[3]{-\frac{q}{2} - \sqrt{\left(\frac{q}{2}\right)^2 + \left(\frac{p}{3}\right)^3}}. \tag{2.37}$$

This equation has three real roots if

$$\left(\frac{q}{2}\right)^2 + \left(\frac{p}{3}\right)^3 < 0.$$

This negative quantity appears under a square root, then a paradox grows that the solution of a geometrical problem is obtained by means of quantity that does not have a geometrical meaning.

Today we say that these solutions are the sum of two complex conjugate quantities, therefore the final result is real.

Raffaele Bombelli, at the end of 16th century has introduced (*Algebra*, Bologna, 1572), complex numbers for the solution of cubic equations, by formulating, practically in modern form, the four operations with complex numbers and introducing the expression that today we write $a + ib$.[6] These numbers have been called *imaginary* by Descartes and this name is also the actual one. The way

---

[6] Bombelli writes: ◁ ... also if this introduction can appear as an extravagant idea and I considered it, for some time, as sophistical rather than true, I have found the demonstration that it works well in the operations. ▷

for the modern formalization has required more than two centuries and has been completed by Euler and Gauss by:

- the introduction of the imaginary unity i,
- to call *complex numbers* the binomial $a + ib$
- to introduce the *functions of a complex variable*, for relevant physical (Euler: the motion of fluids) and geometrical (Gauss: conformal mapping) applications.

2.6.2 Geometrical and Classical Definition of Hyperbolic Angles

2.6.2.1 Geometrical Meaning of Hyperbolic Angle

We have seen that the circular trigonometric functions could be introduced, in a formal way, by means of Euler's formula (2.7). In a similar way the hyperbolic trigonometric functions can be introduced. Actually by summing and subtracting (2.27) and (2.28) for $\rho = 1$ we obtain a formal relation between the hyperbolic trigonometric functions and exponential function of a hyperbolic quantity

$$\cosh\theta \equiv x = \frac{\exp[h\,\theta] + \exp[-h\,\theta]}{2}; \quad \sinh\theta \equiv y = \frac{\exp[h\,\theta] - \exp[-h\,\theta]}{2h}. \quad (2.38)$$

We can check that also from these relations (2.29) follows. From (2.29) we can obtain the geometrical interpretation of hyperbolic trigonometric functions. Actually let us consider in the x, y Cartesian plane, the curve

$$x = \cosh\theta, \quad y = \sinh\theta, \quad (2.39)$$

as function of the parameter θ.

This curve, taking into account that $\cosh\theta > \sinh\theta$ and $\cosh\theta > 1$ represents the right arm of the unitary equilateral hyperbola $x^2 - y^2 = 1$.

Therefore, by analogy with the circular angles defined on the unitary circle, we can call θ the hyperbolic angle and define $\cosh\theta$ and $\sinh\theta$ as the abscissa and the ordinate of the hyperbola point defined by θ, respectively (see Fig. 2.4).

Now we see another correspondence with the circular trigonometric functions. Actually we know that trigonometric angles, measured by radiants, are equal to twice the area of the circular sectors they identify. The same is true for hyperbolic angles. We have

Theorem 2.3 *The hyperbolic angle θ is twice the area of the sector OVP (Fig. 2.4)*

Proof The area of the sector *OVP* is given by the difference between the areas of triangle *OHP* and *VHP*. Therefore, by means of (2.39), we have

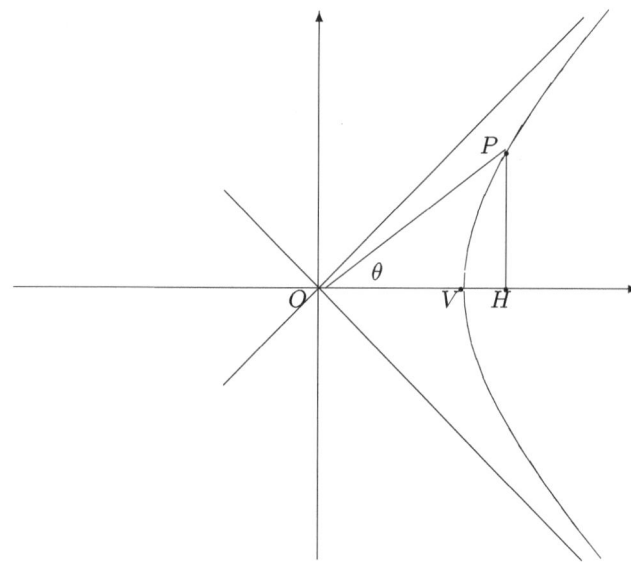

Fig. 2.4 Geometrical definition of hyperbolic angles. The trigonometric circular functions are defined by means of goniometric circle. In a similar way the hyperbolic trigonometric functions can be defined by means of the unitary equilateral hyperbola. Actually let us consider the right arm of hyperbola $x^2 - y^2 = 1$ and define an angle θ corresponding to half-line OP so that $\cosh\theta = \overline{OH}, \sinh\theta = \overline{HP}$. In Appendix 2.6.2 we see that, as for circular angles measured in radiants, also to hyperbolic trigonometric angles θ we can give the geometrical meaning of an area $\theta = 2\mathrm{area}(OVP)$. In Chap. 4, we also see that this area has the same value measured in both "hyperbolic" or "Euclidean" way

$$\mathrm{area}(OHP) = \frac{1}{2}\sinh\theta\cosh\theta = \frac{\sinh 2\theta}{4}$$

$$\mathrm{area}(VHP) = \int_0^\theta y\,dx \equiv \int_0^\theta \sinh^2\theta\,d\theta = \frac{\sinh 2\theta}{4} - \frac{\theta}{2} \qquad (2.40)$$

$$\mathrm{area}(OVP) \equiv \mathrm{area}(OHP) - \mathrm{area}(VHP) = \frac{\theta}{2}.$$

The integral is solved by means of (4.28). ☐

Now we see how this definition allows one to introduce the classical hyperbolic trigonometric functions.

2.6.2.2 Classical Definition of Hyperbolic Trigonometric Functions

Let us consider the equation of hyperbola in Cartesian coordinates $x^2 - y^2 \equiv (x+y)(x-y) = 1$ that can also be written

$$y = \pm\sqrt{x^2 - 1} \tag{2.41}$$

$$x - y = \frac{1}{x + y} \tag{2.42}$$

we have

$$\text{area}(OHP) = \frac{xy}{2} \equiv \frac{x\sqrt{x^2 - 1}}{2} \tag{2.43}$$

$$\text{area}(VHP) = \int_1^x y\,dx \equiv \int_1^x \sqrt{x^2 - 1}\,dx \equiv \frac{x\sqrt{x^2 - 1} - \ln(x + \sqrt{x^2 - 1})}{2}$$
$$\equiv \frac{x\sqrt{x^2 - 1} - \ln(x + y)}{2} \tag{2.44}$$

$$\text{area}(OVP) \equiv \text{area}(OHP) - \text{area}(VHP) = \frac{\ln(x + y)}{2}. \tag{2.45}$$

Comparing (2.40) with (2.45) we have

$$\theta = \ln(x + y) \Rightarrow x + y = \exp[\theta], \tag{2.46}$$

and from (2.42)

$$\ln(x - y) = -\ln(x + y) \Rightarrow \ln(x - y) = -\theta \Rightarrow x - y = \exp[-\theta]. \tag{2.47}$$

Summing and subtracting (2.46) and (2.47), the classical definition follows

$$\cosh\theta \equiv x = \frac{\exp[\theta] + \exp[-\theta]}{2}, \quad \sinh\theta \equiv y = \frac{\exp[\theta] - \exp[-\theta]}{2}. \tag{2.48}$$

References

1. F. Catoni, D. Boccaletti, R. Cannata, V. Catoni, E. Nichelatti, P. Zampetti, *The Mathematics of Minkowski Space–Time* (Birkhäuser Verlag, Basel, 2008)
2. F. Catoni, R. Cannata, V. Catoni, P. Zampetti, N-dimensional geometries generated by hypercomplex numbers. Adv. Appl. Clifford Al. **15**(1), 1 (2005)
3. C.C. Silva, R. de Andrade Martins, Polar and axial vectors versus quaternions. Am. J. Phys. **70**(9), 958 (2002)
4. M. Lavrentiev, B. Chabat, *Effets Hydrodynamiques et modèles mathématiques* (Mir, Moscou, 1980)
5. I.M. Yaglom, *Complex Numbers in Geometry* (Academic Press, New York, 1968)
6. I.M. Yaglom, *A Simple Non-Euclidean Geometry and its Physical Basis* (Springer, New York, 1979)
7. G.L. Naber, *The Geometry of Minkowski Spacetime. An Introduction to the Mathematics of the Special Theory of Relativity*, Sect. 1.4 (Springer, New York, 1992)

Chapter 3
Geometrical Representation of Hyperbolic Numbers

Abstract A relevant property of Euclidean geometry is the Pythagorean distance between two points. From this definition the properties of analytical geometry follow. In a similar way the analytical geometry in Minkowski plane is introduced, starting from the invariant quantities of Special Relativity.

Keywords Distance in Minkowski plane · Distance in space-time · Analytical geometry in space-time

3.1 Introduction

In this chapter we formalize the analytic geometry into hyperbolic plane, i.e., to obtain the *"geometrical consequences" deriving from the definition of "distance" by means of a non-definite quadratic form.*

Let us consider the two-dimensional system of hyperbolic numbers defined as

$$\{z = x + \mathrm{h}\,y; \quad \mathrm{h}^2 = 1; \quad x, y \in \mathbf{R}, \quad \mathrm{h} \notin \mathbf{R}\},$$

and let us introduce a hyperbolic Cartesian plane by analogy with the Gauss–Argand plane of a complex variable.

In this plane we associate points $P \equiv (x, y)$ with hyperbolic numbers $z = x + \mathrm{h}\,y$ and define their quadratic distance from the origin of coordinates as

$$D = z\tilde{z} \equiv x^2 - y^2. \tag{3.1}$$

Therefore the equilateral hyperbolas $x^2 - y^2 = \text{const.}$ represent the locus of points at the same distance from the coordinates origin, as the circles in Euclidean plane. We will see in Chap. 5, that also the theorems, usually stated for circles, hold in hyperbolic plane for equilateral hyperbolas.

The definition of distance (metric element) is equivalent to introducing the bilinear form of the *scalar product*. The scalar product and the properties of hyperbolic numbers allow one to state suitable axioms ([1], p. 245) and to give the structure of a vector space to hyperbolic plane.

F. Catoni et al., *Geometry of Minkowski Space–Time,* SpringerBriefs in Physics,
DOI: 10.1007/978-3-642-17977-8_3, © Francesco Catoni 2011

We begin by seeing the topological properties of hyperbolic plane by considering the multiplicative inverse of z that, if it exists, is given by $1/z \equiv \tilde{z}/z\tilde{z}$. This implies that z does not have an inverse when $z\tilde{z} \equiv x^2 - y^2 = 0$, i.e., when $y = \pm x$, or alternatively when $z = x \pm \mathrm{h} x$. For these numbers, that, as shown in Sect. 2.2, are called "divisors of zero", also the product is not univocally defined.

These numbers are represented on two straight lines, the axes bisectors, that divide the hyperbolic plane in four sectors that we shall call *Right sector (Rs)*, *Up sector (Us)*, *Left sector (Ls)*, and *Down sector (Ds)*.

Now let us consider the quadratic distance (3.1), which is positive in Rs, Ls ($|x| > |y|$) sectors, and negative in Us, Ds ($|x| < |y|$) sectors. As we better see in this and next chapters, this quantity must have its sign and appear in this quadratic form.

It must be noted that this mathematical exposition is more general than the requirements of special relativity where, by giving the physical meaning of time to one of the variable, it must be $t > x$ then $t^2 - x^2 > 0$.

When we must use the linear form (the modulus of hyperbolic numbers or the length of a segment), we follow the definition of Yaglom ([1], p. 180), by taking the absolute value of the quadratic distance

$$\rho = \sqrt{|z\tilde{z}|} \equiv \sqrt{|D|}. \tag{3.2}$$

3.2 Extension of Hyperbolic Exponential Function and Hyperbolic Polar Transformation

We have seen that the exponential function (2.26) allows us to introduce the hyperbolic polar transformation but just for x, y in the Rs sector of hyperbolic plane. This representation can be extended to the complete (x, y) plane, by means of the transformations

$$\text{if } |x| > |y|; \quad x + \mathrm{h}\,y = \text{sign}(x)\rho \exp[\mathrm{h}\,\theta] \equiv \text{sign}(x)\rho(\cosh\theta + \mathrm{h}\,\sinh\theta); \tag{3.3}$$

$$\text{if } |x| < |y|; +\mathrm{h}\,y = \text{sign}(y)\rho\,\mathrm{h}\exp[\mathrm{h}\,\theta] \equiv \text{sign}(y)\rho(\sinh\theta + \mathrm{h}\,\cosh\theta), \tag{3.4}$$

where $\rho = \sqrt{|x^2 - y^2|}$ and θ is defined by

$$\text{for } |x| > |y|, \quad \theta = \tanh^{-1}(y/x); \quad \text{for } |x| < |y|, \quad \theta = \tanh^{-1}(x/y),$$

This extension is reported in Table 3.1.

Table 3.1 Map of the complete (x, y) plane by hyperbolic polar transformation

$|x| > |y|$		$|x| < |y|$	
Right sector (Rs)	Left sector (Ls)	Up sector (Us)	Down sector (Ds)
$z = \rho \exp[\mathrm{h}\,\theta]$	$z = -\rho \exp[\mathrm{h}\,\theta]$	$z = \mathrm{h}\,\rho \exp[\mathrm{h}\,\theta]$	$z = -\mathrm{h}\,\rho \exp[\mathrm{h}\,\theta]$
$x = \rho \cosh\theta$	$x = -\rho \cosh\theta$	$x = \rho \sinh\theta$	$x = -\rho \sinh\theta$
$y = \rho \sinh\theta$	$y = -\rho \sinh\theta$	$y = \rho \cosh\theta$	$y = -\rho \cosh\theta$

3.3 Geometry in the Hyperbolic Plane

Now we restate for the hyperbolic Cartesian plane some classical definitions and properties of the Euclidean Cartesian plane.

- Definitions
  Given two points $P_j \equiv (x_j, y_j)$, $P_k \equiv (x_k, y_k)$ that are associated with the hyperbolic variables $z_j = x_j + \mathrm{h}\, y_j$ and $z_k = x_k + \mathrm{h}\, y_k$, we define the quadratic distance between them by extending (3.1),

$$D_{jk} = (z_j - z_k)(\tilde{z}_j - \tilde{z}_k). \tag{3.5}$$

As a general rule we indicate the quadratic segment lengths by capital letters, and by the same small letters the square root of their absolute value

$$d_{jk} = \sqrt{|D_{jk}|}. \tag{3.6}$$

Following ([1], p. 179) a segment or line is said to be of the *first (second) kind* if it is parallel to a line through the origin located in the sectors containing the $x(y)$ axis. This classification of straight lines is equivalent to the one of special relativity recalled in Sect. 2.3. Actually if one variable has the physical meaning of time, in the t, x plane, we have

1. the straight lines called of the *first kind* are the same as the time-like,
2. the straight lines called of the *second kind* are the same as the space-like.

   Therefore the segment $\overline{P_j P_k}$ is of the first (second) kind if $D_{jk} > 0$ ($D_{jk} < 0$).
   We also call for hyperbolic numbers, as it is usual for complex variables, $\Re\{*\}$ the real part and $\mathcal{H}\{*\}$ the coefficient of the hyperbolic versor.

- Straight line equations
  We know that in the Euclidean Cartesian plane the equations of straight lines in normal form, are expressed by means of the circular trigonometric functions. Now we see that the most appropriate expressions of the equations of a straight line in the hyperbolic plane is by means of hyperbolic trigonometric functions. Actually these expressions state the kind of straight lines and reflect the topological characteristics of the hyperbolic plane.

  In particular, depending on their kind, we have two different expressions for the equations of straight lines

  – for $|x - x_0| > |y - y_0|$ (a straight line of the first kind) is written as

$$(x - x_0) \sinh \theta - (y - y_0) \cosh \theta = 0, \tag{3.7}$$

  – for $|x - x_0| < |y - y_0|$ (a straight line of the second kind) is written as

$$(x - x_0) \cosh \theta' - (y - y_0) \sinh \theta' = 0. \tag{3.8}$$

The angle θ' is measured referring to the y axis as explained in Table 4.1.

- Pseudo-orthogonality

Theorem 3.1 *Two straight lines are pseudo-orthogonal if they are symmetric with respect to the straight lines from their crossing point, parallel to axes bisectors*

Proof As for Euclidean plane, two straight lines in the hyperbolic plane are said to be pseudo-orthogonal when the scalar product of their unity vectors (direction cosine) is zero. The definition of scalar product derives from the definition of distance. In particular, as we better see in Sect. 4.1, given two vectors written in the formalism of hyperbolic numbers as

$$v_1 = x_1 + h\,y_1 \quad \text{and} \quad v_2 = x_2 + h\,y_2,$$

the scalar product is defined as

$$\Re(v_1 \tilde{v}_2) \equiv \Re[(x_1 + h\,y_1)(x_2 - h\,y_2)] = x_1\,x_2 - y_1\,y_2. \tag{3.9}$$

Let us consider two straight lines of the first kind v_1, v_2 expressed by (3.7), with angular coefficients given by the hyperbolic angles θ_1, θ_2, the direction cosines will be: for $v_1 \equiv (\sinh\theta_1, \cosh\theta_1)$ and for $v_2 \equiv (\sinh\theta_2, \cosh\theta_2)$. Their scalar product is

$$v_1 \cdot v_2 = \sinh\theta_1 \sinh\theta_2 - \cosh\theta_1 \cosh\theta_2 \equiv -\cosh(\theta_1 - \theta_2) \neq 0 \tag{3.10}$$

Therefore two straight lines of the same kind, cannot be pseudo-orthogonal.

Now, together with v_1, we consider the straight line of the second kind $v_2' \equiv (\cosh\theta_2', \sinh\theta_2')$. Their scalar product is:

$$v_1 \cdot v_2' = \sinh\theta_1 \cosh\theta_2' - \cosh\theta_1 \sinh\theta_2' \equiv \sinh(\theta_1 - \theta_2'). \tag{3.11}$$

Then a straight line of the first kind (3.7) has a pseudo-orthogonal line of the second kind (3.8) with the same angle ($\theta_1 = \theta_2'$), and conversely.

Taking into account (Fig. 3.1) that θ_1 is measured with respect to x axis and θ_2' with respect to y axis, we have the demonstration of theorem. $\square$

This result, well known in special relativity, is represented in Fig. 3.2

Let us now see some properties similar to straight lines in Euclidean and hyperbolic planes.

- It is known that in complex formalism, the equation of a straight line is given by

$$\Re[(x + i\,y)(\exp[i\,\phi]) + A + iB] = 0; \quad A, B \in \mathbf{R}, \tag{3.12}$$

and its orthogonal straight line by

$$\Im[(x + i\,y)(\exp[i\,\phi]) + A + iB] = 0; \quad A, B \in \mathbf{R}.$$

The same result holds in the hyperbolic plane in hyperbolic formalism.

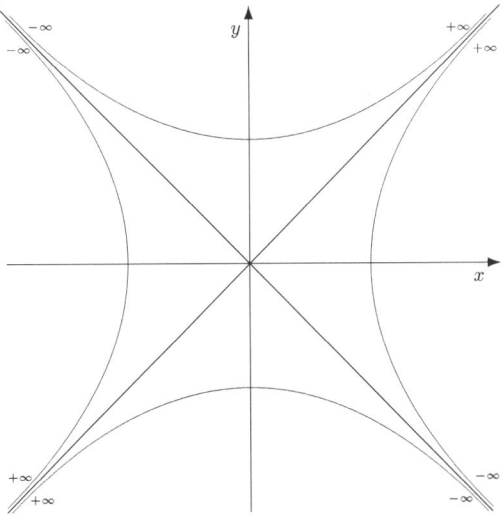

Fig. 3.1 Indication of direction paths on the four arms of hyperbola as the parameter θ goes from $-\infty$ to $+\infty$. For $\rho = 1$ the x, y in Table 3.1 represent the four arms of equilateral hyperbolas $|x^2 - y^2| = 1$. These four arms of hyperbola correspond, in the hyperbolic plane, to goniometric circle that in Euclidean plane is used for defining the trigonometric functions. In Sect. 2.6.2, we have seen that we can give to θ the meaning of a "hyperbolic angle" and allows one to define the hyperbolic trigonometric functions in Rs. In Table 3.1 we have extended the polar hyperbolic transformation in the complete x, y plane, in the same way we now extend the hyperbolic trigonometric functions in the whole plane, by defining them on the four arms of equilateral hyperbola $|x^2 - y^2| = 1$ and measuring θ with respect to the coordinate semi-axis of the corresponding sector. In Sect. 4.3 (Figs. 4.1 and 4.2), we see the analytical and geometrical formalization of this extension. We can note the different symmetry from Euclidean and pseudo-Euclidean planes: in the former the angles in all sectors are increasing in the anticlockwise direction; for the latter the sign of the angles is symmetric with respect to axes bisectors. Therefore in sectors Us and Ds they are increasing clockwise

Actually we can write the equation of a straight line in a hyperbolic plane as

$$\Re[(x + h\,y)(\exp[h\,\theta]) + A + h\,B] = 0; \quad A, B \in \mathbf{R}. \tag{3.13}$$

We can check that the hyperbolic part

$$\mathcal{H}[(x + h\,y)(\exp[h\,\theta]) + A + h\,B] = 0; \quad A, B \in \mathbf{R} \tag{3.14}$$

represents its pseudo-orthogonal straight line. We note that the product of the angular coefficients for two pseudo-orthogonal lines is $+1$, instead of -1 as in the Euclidean plane.

– For complex numbers the Euler's formula (2.3) gives

$$i = \exp[i\,\pi/2]. \tag{3.15}$$

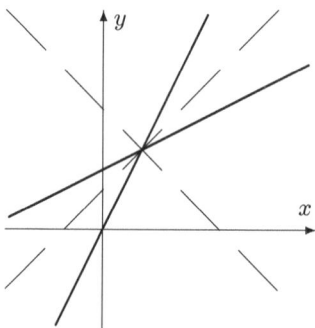

Fig. 3.2 Pseudo-orthogonal straight lines. From an analytical point of view two straight lines are orthogonal if the scalar product of the direction cosines is null. This definition is extended to pseudo-Euclidean plane. As we can see in Sect. 4.2, the definition of the scalar product depends on the definition of distance, then its expression in hyperbolic plane is different from the Euclidean scalar product. As it is shown by Theorem 3.1, in a Euclidean representation two straight lines are pseudo-orthogonal if they are symmetric with respect to the parallels to the axes bisectors from their crossing point

Therefore to multiply a vector for i, is equivalent to rotate it by $\pi/2$. Otherwise we have just recalled that the real and the imaginary part of the straight line (3.12) represent two orthogonal straight lines both in the complex and hyperbolic planes. Actually we see in next chapter that it is possible and important to extend the hyperbolic trigonometric functions. This extension allows one to state a relation equivalent with (3.15).

Now we show this property with a simple test. Actually we have $h(x + hy) = y + hx$, therefore the multiplication by h gives a point symmetric with respect to axes bisectors, and the straight lines from the origin to symmetric points are pseudo-orthogonal.

- Axis of a segment
  As in the Euclidean plane, we have

Theorem 3.2 *The axis of a segment in hyperbolic plane is pseudo-orthogonal to a segment in its middle point.*

Proof Let us consider two points $P_1 \equiv (x_1, y_1)$, $P_2 \equiv (x_2, y_2)$. The points that have the same hyperbolic distance from these two points are determined by equation

$$\overline{PP_1}^2 = \overline{PP_2}^2 \Rightarrow (x - x_1)^2 - (y - y_1)^2 = (x - x_2)^2 - (y - y_2)^2$$
$$\Rightarrow (x_1 - x_2)(2x - x_1 - x_2) = (y_1 - y_2)(2y - y_1 - y_2) \qquad (3.16)$$

and, in canonical form,

$$y = \frac{(x_1 - x_2)}{(y_1 - y_2)} x + \frac{(y_1^2 - y_2^2) - (x_1^2 - x_2^2)}{2(y_1 - y_2)}. \qquad (3.17)$$

Therefore from (3.17) it follows that the axis is pseudo-orthogonal to segment $\overline{P_1 P_2}$, and from (3.16) that it passes through its middle point $P_M \equiv ((x_1 + x_2)/2, (y_1 + y_2)/2)$. □

- Distance of a point from a straight line

Theorem 3.3 *The distance of a point P_1 from a straight line γ is proportional to the result of substituting the coordinates of P_1 in the equation for γ.*

Proof Let us take a point $P \equiv (x, y)$ on the straight line $\gamma : \{y - mx - q = 0\}$, and a point $P_1 \equiv (x_1, y_1)$ outside the straight line. The quadratic distance

$$\overline{PP_1}^2 = (x - x_1)^2 - (y - y_1)^2$$

has its extreme for the point P_2 of the straight line, with abscissa

$$x \equiv x_2 = (x_1 - my_1 - mq)/(1 - m^2),$$

and quadratic distance

$$\overline{P_2 P_1}^2 \equiv D_{12} = \frac{(y_1 - mx_1 - q)^2}{m^2 - 1} \quad \text{and} \quad d_{12} = \frac{|y_1 - mx_1 - q|}{\sqrt{|m^2 - 1|}}. \tag{3.18}$$

It is easy to verify that this distance corresponds to a *maximum* as it is well known from special relativity [2]. From expressions (3.18) the theorem follows. □

The equation of the straight line through P_1 and P_2 is

$$(y - y_1) = \frac{1}{m}(x - x_1),$$

that represents a straight line *pseudo-orthogonal* to γ.

In particular if the equation of the straight line is in the form (3.7) or (3.8), the linear distance is obtained, as in Euclidean geometry, by substituting the point coordinates in the equation of the straight line

$$d_{12} = |(x_1 - x_0)\sinh\theta - (y_1 - y_0)\cosh\theta| \tag{3.19}$$

$$d_{12} = |(x_1 - x_0)\cosh\theta' - (y_1 - y_0)\sinh\theta'| \tag{3.20}$$

References

1. I.M. Yaglom, *A Simple Non-Euclidean Geometry and Its Physical Basis.* (Springer-Verlag, New York, 1979)
2. G.L. Naber, *The Geometry of Minkowski Spacetime. An Introduction to the Mathematics of the Special Theory of Relativity*, Sect. 1.4 (Springer-Verlag, New York, 1992)

Chapter 4
Trigonometry in the Hyperbolic (Minkowski) Plane

Abstract The correspondence between properties of complex numbers and Euclidean geometry allows to obtain an algebraic formalization of Euclidean geometry. Thanks to the equivalent properties between complex and hyperbolic numbers, the geometry of Minkowski space-time can be formalized in a similar algebraic way. Moreover, introducing two invariant quantities, the complete formalization of space-time trigonometry is obtained.

Keywords Geometry in hyperbolic plane · Space-time invariants · Space-time trigonometry · Theorems on hyperbolic triangles · Hyperbolic triangles solution

4.1 Analytical Formalization of Euclidean Trigonometry

Now we see how it is possible to formalize Euclidean geometry and trigonometry in an algebraic way by means of complex numbers.

Actually, let us consider the geometrical figures represented in a Cartesian plane, we can say, in group theory language (see Appendix A), that Euclidean geometry studies the invariant properties of the geometrical figures under their rotations and translations. By expressing these properties by complex numbers, we see how it is possible to formalize, in a Cartesian plane, the trigonometric functions and to obtain all the trigonometry theorems by an analytical method as mathematical identities [1].

This result is extended to the space–time geometry associated with hyperbolic numbers and, by demonstrating theorems with an algebraic approach, we obtain a complete formalization of space–time geometry and trigonometry.

4.1.1 Complex Numbers Invariants in Euclidean Plane

Let us consider the Gauss–Argand complex plane (Fig. 2.1) where a vector from the origin O to point $P \equiv (x, y)$ is represented by $v = x + \mathrm{i}y$. With this formalism a rotation of a vector of an angle α transforms its components by means of (2.10) that can be written $v' \equiv v \exp[i\alpha]$. We have

$$|v'|^2 \equiv v'\bar{v}' = v \exp[i\alpha]\bar{v}\exp[-i\alpha] = v\bar{v} \equiv |v|^2. \tag{4.1}$$

In a similar way we check that there are two more invariants related to any pair of vectors. Let us consider two vectors from the origin O to point $P_1 \equiv (x_1, y_1)$ and $P_2 \equiv (x_2, y_2)$: $v_1 = x_1 + iy_1 \equiv \rho_1 \exp[i\phi_1]$, $v_2 = x_2 + iy_2 \equiv \rho_2 \exp[i\phi_2]$: we have

Theorem 4.1 *The real and imaginary parts of the product $v_2\bar{v}_1$ are invariant under axes rotations and these two invariants allow us an operative definition of trigonometric functions by means of the components of the vectors:*

$$\cos(\phi_2 - \phi_1) = \frac{x_1x_2 + y_1y_2}{\rho_1\rho_2}; \quad \sin(\phi_2 - \phi_1) = \frac{x_1y_2 - x_2y_1}{\rho_1\rho_2}. \tag{4.2}$$

Proof Actually

$$v_2'\bar{v}_1' = v_2 \exp[i\alpha]\bar{v}_1 \exp[-i\alpha] \equiv v_2\bar{v}_1, \tag{4.3}$$

and let us represent the two vectors in polar coordinates $v_1 \equiv \rho_1 \exp[i\phi_1]$, $v_2 \equiv \rho_2 \exp[i\phi_2]$. $\qquad\square$

Consequently we have

$$v_2\bar{v}_1 = \rho_1\rho_2 \exp[i(\phi_2 - \phi_1)] \equiv \rho_1\rho_2[\cos(\phi_2 - \phi_1) + i\sin(\phi_2 - \phi_1)]. \tag{4.4}$$

As is well known, the resulting real part of this product represents the scalar product, while the imaginary part represents the area of the parallelogram defined by v_1 and v_2, that represents the modulus of the cross product, i.e., we have obtained just *by means of mathematical considerations the two relevant physical invariants.*
In Cartesian coordinates we have

$$v_2\bar{v}_1 = (x_2 + iy_2)(x_1 - iy_1) \equiv x_1x_2 + y_1y_2 + i(x_1y_2 - x_2y_1), \tag{4.5}$$

and, by comparing (4.4) with (4.5), we obtain (4.2). $\qquad\square$

We know that the theorems of Euclidean trigonometry are usually obtained following a geometric approach; now we can state

Theorem 4.2 *Using the Cartesian expressions of trigonometric functions, given by (4.2), the trigonometry theorems are simple identities.*

Proof We know that the trigonometry theorems represent relations between angles and side lengths of a triangle. If we represent a triangle in a Cartesian plane (Fig. 4.3) it is defined by the coordinates of its vertexes P_1, P_2, P_3. From the coordinates of these points we obtain the side lengths by Pythagoras' theorem and the trigonometric functions from (4.2). By these definitions we can verify that the trigonometry theorems are identities, as we see in the application of this method to hyperbolic trigonometry. $\qquad\square$

Now, by extending the exposed procedure to hyperbolic plane, we formalize the hyperbolic trigonometry.

4.2 Hyperbolic Rotation Invariants in Minkowski Plane

By analogy with the Euclidean trigonometry approach, just summarized, we can say that *pseudo-Euclidean plane geometry* studies the properties that are invariant under two-dimensional Lorentz transformations (Lorentz–Poincaré group of special relativity) corresponding to hyperbolic rotation (Sect. 2.4.1). We show afterward, how these properties allow us to formalize hyperbolic trigonometry.

Let us define in the hyperbolic plane (Fig. 2.3), a hyperbolic vector, from the origin to point $P \equiv (x, y)$, as $v = x + hy$.

A hyperbolic rotation of an angle θ_a transforms the components of this vector by means of (2.31), that can be written $v' = v \exp[h\theta_a]$ and we verify the invariance of the modulus in the hyperbolic axes rotations

$$|v'|^2 \equiv v'\tilde{v}' = v \exp[h\theta_a]\tilde{v} \exp[-h\theta_a] \equiv v\tilde{v} = |v|^2 \tag{4.6}$$

In a similar way we check that, as for complex numbers, there are two more invariants related to any pair of vectors.

Let us consider two vectors from the origin O to points $P_1 \equiv (x_1, y_1)$ and $P_2 \equiv (x_2, y_2)$: $v_1 = x_1 + hy_1$ and $v_2 = x_2 + hy_2$, we have

Theorem 4.3 *The real and the hyperbolic parts of the product $v_2\tilde{v}_1$ are invariant under hyperbolic rotation, and these two invariants allow us an operative definition of hyperbolic trigonometric functions by means of the components of the vectors*:

$$\cosh(\theta_2 - \theta_1) = \frac{x_1 x_2 - y_1 y_2}{\rho_1 \rho_2} \equiv \frac{x_1 x_2 - y_1 y_2}{\sqrt{(x_2^2 - y_2^2)(x_1^2 - y_1^2)}}, \tag{4.7}$$

$$\sinh(\theta_2 - \theta_1) = \frac{x_1 y_2 - x_2 y_1}{\rho_1 \rho_2} \equiv \frac{x_1 y_2 - x_2 y_1}{\sqrt{(x_2^2 - y_2^2)(x_1^2 - y_1^2)}} \tag{4.8}$$

Proof We have

$$v_2'\tilde{v}_1' = v_2 \exp[h\theta_a]\tilde{v}_1 \exp[-h\theta_a] \equiv v_2\tilde{v}_1. \tag{4.9}$$

Let us suppose $(x, y) \in R_s$, and represent the two vectors in hyperbolic polar form $v_1 = \rho_1 \exp[h\theta_1]$, $v_2 = \rho_2 \exp[h\theta_2]$. Consequently we have

$$v_2\tilde{v}_1 \equiv \rho_1 \rho_2 \exp[h(\theta_2 - \theta_1)] \equiv \rho_1 \rho_2[\cosh(\theta_2 - \theta_1) + h\sinh(\theta_2 - \theta_1)]. \tag{4.10}$$

In Cartesian coordinates, we have

$$v_2\tilde{v}_1 = (x_2 + hy_2)(x_1 - hy_1) \equiv x_1 x_2 - y_1 y_2 + h(x_1 y_2 - x_2 y_1). \tag{4.11}$$

Comparing (4.10) with (4.11) we obtain (4.7) and (4.8). □

This result is the same obtained for complex numbers. Actually the real term of (4.11) is a bilinear expression that as Cartesian expression of the scalar product depends on the definition of "distance": in the hyperbolic plane, due to the different sign in the distance definition (3.1), we have a different sign with respect to Euclidean scalar product.

As far as the hyperbolic part is concerned, we shall see in Sect. 4.5.1 that, as for the Euclidean plane, it represents a *pseudo-Euclidean area*.

4.3 Extension of Hyperbolic Trigonometric Functions

All points of Euclidean Cartesian plane x, y can be represented by means of the polar transformation $x = \rho \cos \phi, y = \rho \sin \phi; \rho > 0, 0 \le \phi < 2\pi$. A similar transformation performed by substituting the hyperbolic to the circular trigonometric functions, represents just points $x > 0, -x < y < x$ (sector *Rs*). Therefore for representing all the points we need the four functions of Table 3.1. Now we show

Theorem 4.4 *Equations 4.7 and 4.8 allow us to extend the hyperbolic trigonometric functions in the complete (x, y) plane.*

Proof If we set $v_1 \equiv (1, 0)$ and $(\theta_2, x_2, y_2) \rightarrow (\theta, x, y)$, (4.7) and (4.8) become

$$\cosh \theta = \frac{x}{\sqrt{|x^2 - y^2|}}, \quad \sinh \theta = \frac{y}{\sqrt{|x^2 - y^2|}}, \tag{4.12}$$

we observe that the expressions in the right-hand sides of (4.12) are valid for $\{x, y \in \mathbf{R} | x \ne \pm y\}$ so they allow us to define the trigonometric hyperbolic functions in the complete (x, y) plane. This extension is the same as the one proposed in [2], that we summarize in Sect. 4.3.1, and will allow us to define the sum of angles also if they are in different sectors. □

In the following we will denote with $\cosh_e, \sinh_e$ these *extended hyperbolic functions*. In Table 4.1 the relations between $\cosh_e, \sinh_e$ and traditional hyperbolic functions are reported. By this extension the hyperbolic polar transformation, (2.26), can be written in terms of just one expression holding in the complete hyperbolic plane

Table 4.1 Relations between functions $\cosh_e, \sinh_e$ and classical hyperbolic functions

	$|x| > |y|$		$|x| < |y|$	
	$(Rs), x > 0$	$(Ls), x < 0$	$(Us), y > 0$	$(Ds), y < 0$
$\cosh_e \theta =$	$\cosh \theta$	$-\cosh \theta$	$\sinh \theta$	$-\sinh \theta$
$\sinh_e \theta =$	$\sinh \theta$	$-\sinh \theta$	$\cosh \theta$	$-\cosh \theta$

From the definition of hyperbolic trigonometric functions by means of two vectors from the coordinate origin, we formalize the hyperbolic trigonometric function in the whole hyperbolic plane. The hyperbolic angle θ in the last four columns is calculated referring to semi-axes $x, -x, y, -y$, respectively and increases as shown in Fig. 3.1

$$x + hy \Rightarrow \rho(\cosh_e \theta + h \sinh_e \theta), \tag{4.13}$$

from which, for $\rho = 1$, we obtain the *extended hyperbolic Euler's formula* [2]

$$\exp_e[h\theta] = \cosh_e \theta + h \sinh_e \theta. \tag{4.14}$$

From Table 4.1 or (4.12) it follows that

$$\begin{aligned} &\text{for } |x| > |y| \Rightarrow \cosh_e^2 \theta - \sinh_e^2 \theta = 1; \\ &\text{for } |x| < |y| \Rightarrow \cosh_e^2 \theta - \sinh_e^2 \theta = -1. \end{aligned} \tag{4.15}$$

Now we see that the extended hyperbolic trigonometric functions can be related to circular trigonometric functions. Actually by giving to x, y in (4.12) all the values on the circle $x = \cos \phi, y = \sin \phi$ for $0 \le \phi < 2\pi$, we obtain

$$\begin{aligned} \cosh_e \theta &= \frac{\cos \phi}{\sqrt{|\cos 2\phi|}} \equiv \frac{1}{\sqrt{|1 - \tan^2 \phi|}}; \\ \sinh_e \theta &= \frac{\sin \phi}{\sqrt{|\cos 2\phi|}} \equiv \frac{\tan \phi}{\sqrt{|1 - \tan^2 \phi|}}. \end{aligned} \tag{4.16}$$

These expressions have a simple geometrical interpretation. We have

Theorem 4.5 *Equations* 4.16 *represent a bijective mapping between points on a unit circle (specified by ϕ) and points on a unit hyperbola (specified by θ). From a geometrical point of view represent the projection of the points of the unit circle on the points of unit hyperbola, from the coordinate origin.*

Proof Referring to Fig. 4.1, let us consider the half-line $y = x \tan \phi, x > 0$ which crosses the unit circle, with center $O \equiv (0,0)$, in $P_C \equiv (\cos \phi, \sin \phi)$. The half-line crosses the unit hyperbola with center O at point

$$P'_{\mathcal{I}} \equiv \left(\frac{1}{\sqrt{1 - \tan^2 \phi}}, \frac{\tan \phi}{\sqrt{1 - \tan^2 \phi}} \right) \in Rs \quad \text{for } |\tan \phi| < 1 \tag{4.17}$$

or at point

$$P''_{\mathcal{I}} \equiv \left(\frac{1}{\sqrt{\tan^2 \phi - 1}}; \frac{\tan \phi}{\sqrt{\tan^2 \phi - 1}} \right) \in Us, Ds \quad \text{for } |\tan \phi| > 1. \tag{4.18}$$

The half-line $y = x \tan \phi, x < 0$ crosses the left side of the circle and one of the arms Ls, Us, Ds of the hyperbola. Since the points of unit hyperbolas are given by $P_{\mathcal{I}} \equiv (\cosh_e \theta; \sinh_e \theta)$, by comparing (4.17) and (4.18) with the last terms of (4.16), we have the assertion. $\qquad\square$

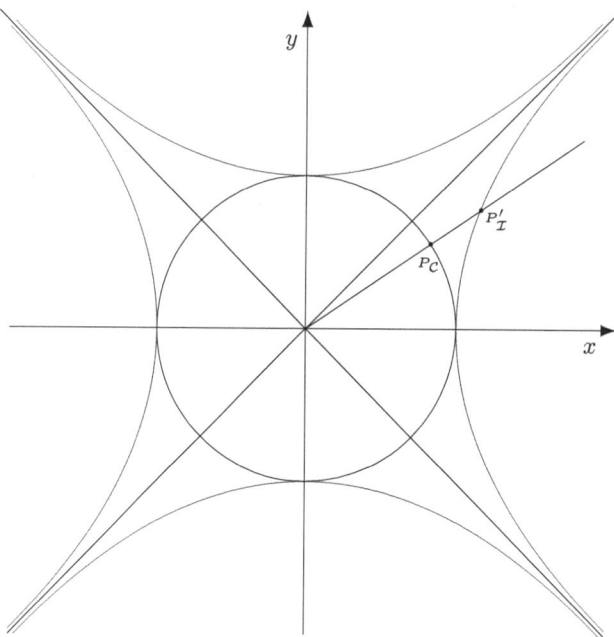

Fig. 4.1 Geometrical correspondence between circular and hyperbolic trigonometric functions. The hyperbolic trigonometric functions can be related to circular trigonometric functions by means of the relations (4.16). As it is demonstrated by Theorem 4.5, these relations have also a simple geometrical interpretation: actually they represent the coordinates of the points of the unitary circle with center in the coordinates origin, projected into the four arms of unitary equilateral hyperbolas

A graph of the function $\cosh_e$ as function of the circular angle ϕ from (4.16) is shown in Fig. 4.2.

The fact that the extended hyperbolic trigonometric functions can be represented in terms of just one expression given by (4.12) allows a direct application of these functions for the solution of triangles with sides in any direction, except the directions parallel to axes bisectors.

4.3.1 Fjelstad's Extension of Hyperbolic Trigonometric Functions

In the complex Gauss–Argand plane, the goniometric circle used for the definition of trigonometric functions is expressed by $x + \mathrm{i}y = \exp[\mathrm{i}\phi]$. In the hyperbolic plane the hyperbolic trigonometric functions can be defined on the unit equilateral

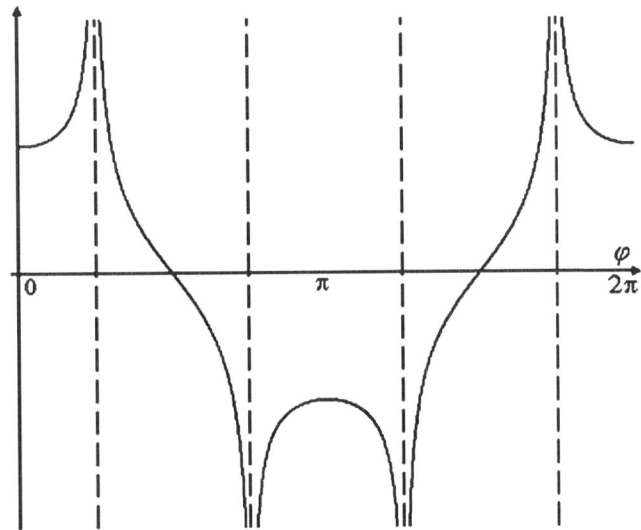

Fig. 4.2 Graphic representation of the function $\cosh_e \theta = \frac{\cos\phi}{\sqrt{|\cos 2\phi|}}$ for $0 \leq \phi < 2\pi$. All points of Cartesian plane x, y can be represented by means of the polar transformation $x = \rho \cos\phi, y = \rho \sin\phi, \rho > 0, 0 < \phi < 2\pi$. A similar transformation performed by substituting the hyperbolic to the circular trigonometric functions, represents just points $x > 0, -x < y < x$ (sector Rs). Therefore for representing all the points we need the four functions of Table 3.1. In Sect. 4.3 we have shown that the hyperbolic trigonometric functions can be set as a bijective correspondence with the circular trigonometric functions (4.16). In this way they can be extended and represent all points of x, y plane by means of just one function. In this figure we represent the function $\cosh_e \theta = \frac{\cos\phi}{\sqrt{|\cos 2\phi|}}$ for $0 \leq \phi < 2\pi$. The broken vertical lines represent the values for which $\cos 2\phi = 0 \Rightarrow x = \pm y$. Because $\sin\phi = \cos(\phi - \pi/2)$ and $|\cos 2\phi| = |\cos(2\phi \pm \pi)|$, from (4.16) we have that the function $\sinh_e \theta$ has the same behavior of $\cosh_e \theta$ allowing for a shift of $\pi/2$. In particular $\sinh_e \theta > 0$ in the range of ϕ values $0 < \phi < \pi$

hyperbola, which can be expressed in a way similar to the goniometric circle: $x + hy = \exp[h\theta]$. However this expression represents only the arm of unit equilateral hyperbolas $\in Rs$. If we want to extend the hyperbolic trigonometric functions on the whole plane, we must take into account all arms of the unit equilateral hyperbola $|x^2 - y^2| = 1$. As we can obtain from (3.3) and (3.4), these arms are given by $x + hy = \pm \exp[h\theta]$ and $x + hy = \pm h \exp[h\theta]$.

Here we summarize the approach followed in [2] which demonstrates how these unit curves allow us to extend the hyperbolic trigonometric functions and to obtain the addition formula for angles in any sector.

These unit curves are the set of points U, where $U = \{z | \rho(z) = 1\}$. Clearly $U(\cdot)$ is a group.

For $z \in Rs$ the group $U(\cdot)$ is isomorphic to $\theta(+)$ where $\theta \in \mathbf{R}$ is the angular function that for $-\infty < \theta < \infty$ traverses the arm $\in Rs$ of the unit hyperbolas. Now we can have a complete isomorphism between $U(\cdot)$ and the angular function, extending the last one to other sectors. This can be done by Klein's four-group $k \in K = \{1, \mathrm{h}, -1, -\mathrm{h}\}$.

Indeed let us consider the expressions of the four arms of the hyperbolas (Table 3.1 for $\rho = 1$). We can extend the angular functions as a product of $\exp[\mathrm{h}\theta]$ and Klein's group, writing $U = \{k \exp[\mathrm{h}\theta] | \theta \in \mathbf{R}, k \in K\}$. Calling $U_k = \{k \exp[\mathrm{h}\theta] | \theta \in \mathbf{R}\}$, the hyperbola arm with the value k and, in the same way θ_k the ordered pair (θ, k), we define $\Theta \equiv \mathbf{R} \times K = \{\theta_k | \theta \in \mathbf{R}, k \in K\}$ and $\Theta_k \equiv \mathbf{R} \times \{k\} = \{\theta_k | \theta \in \mathbf{R}\}$. Θ_1 is isomorphic to $\mathbf{R}(+)$; therefore, accordingly, we think of $\Theta(+)$ as an extension of $\mathbf{R}(+)$. To define the complete isomorphism between $\Theta(+)$ and $U(\cdot)$, we have to define the addition rule for angles: $\theta_k + \theta'_{k'}$. This rule is obtained from the isomorphism itself,

$$\theta_k + \theta'_{k'} \Rightarrow U_k \cdot U_{k'} \equiv k \exp[\mathrm{h}\theta] \cdot k' \exp[\mathrm{h}\theta'] \equiv kk' \exp[\mathrm{h}(\theta + \theta')] \Rightarrow (\theta + \theta')_{kk'}.$$
(4.19)

On this basis we can obtain the hyperbolic angle θ and the Klein index (k) if we know the extended hyperbolic trigonometric functions $\sinh_e \theta$ and $\cosh_e \theta$. We have

$$\text{if } |\sinh_e \theta| < |\cosh_e \theta| \Rightarrow \theta = \tanh^{-1}\left(\frac{\sinh_e \theta}{\cosh_e \theta}\right), \quad k = \frac{\cosh_e \theta}{|\cosh_e \theta|} \cdot 1;$$

$$\text{if } |\sinh_e \theta| > |\cosh_e \theta| \Rightarrow \theta = \tanh^{-1}\left(\frac{\cosh_e \theta}{\sinh_e \theta}\right), \quad k = \frac{\sinh_e \theta}{|\sinh_e \theta|} \cdot \mathrm{h}.$$
(4.20)

4.3.1.1 Application of "Klein's Index" to the Euclidean Plane

Now we see how this "extension" of hyperbolic trigonometric functions can be applied to circular angles giving well-known results.

By analogy with hyperbolic trigonometric functions we define the circular trigonometric functions just in sector Ls, i.e., for $-\pi/4 < \phi < \pi/4$, by means of Euler's formula $\cos \phi + \mathrm{i} \sin \phi = \exp[\mathrm{i}\phi]$. Let us consider the product $k \exp[\mathrm{i}\phi]$ where $k \in K = \{1, \mathrm{i}, -1, -\mathrm{i}\}$ is a four-value group that for $-\pi/4 < \phi < \pi/4$ allows us to obtain the complete circle. The meaning of this product is well known, $\mathrm{i} \exp[\mathrm{i}\phi] \equiv \exp[\mathrm{i}(\frac{\pi}{2} + \phi)]; - \exp[\mathrm{i}\phi] \equiv \exp[\mathrm{i}(\pi + \phi)]; -\mathrm{i} \exp[\mathrm{i}\phi] \equiv \exp[\mathrm{i}(\frac{3\pi}{2} + \phi)]$.

These expressions allow one to clarify the properties of circular trigonometric functions which are determined on the whole circle, by their values for $0 < \phi < \pi/4$.

4.4 Goniometry in the Minkowski Plane

The following expressions are obtained in Euclidean goniometry, by means of geometrical observations, also if they can be obtained by means of (2.3). For hyperbolic goniometry they are obtained by means of (2.27).

- Hyperbolic angles addition formulas.
  Extending to hyperbolic exponential the properties of real and complex exponential, we have

$$\exp[h(\alpha \pm \beta)] = \exp[h\alpha]\ \exp[\pm h\beta],$$

and, applying to both sides the hyperbolic Euler formula (2.27), we have

$$\cosh(\alpha \pm \beta) + h \sinh(\alpha \pm \beta) = (\cosh \alpha + h \sinh \alpha)(\cosh \beta \pm h \sinh \beta). \quad (4.21)$$

Equating the coefficients of the same versors, we obtain the hyperbolic trigonometric functions of the sum and difference of hyperbolic angles

$$\cosh(\alpha \pm \beta) = \cosh \alpha \cosh \beta \pm \sinh \alpha \sinh \beta, \quad (4.22)$$

$$\sinh(\alpha \pm \beta) = \sinh \alpha \cosh \beta \pm \cosh \alpha \sinh \beta. \quad (4.23)$$

From their ratio, after reduction, we obtain

$$\tanh(\alpha \pm \beta) = \frac{\tanh \alpha \pm \tanh \beta}{1 \pm \tanh \alpha \tan \beta}. \quad (4.24)$$

- The double and half angles formulas.
  Setting $\alpha = \beta$ in (4.22) and (4.23), we have

$$\cosh 2\alpha = \cosh^2 \alpha + \sinh^2 \alpha, \quad (4.25)$$

$$\sinh 2\alpha = 2\sinh \alpha \cosh \alpha. \quad (4.26)$$

From (4.25), by means of (4.15), we have

$$\cosh \alpha = \sqrt{\frac{\cosh 2\alpha + 1}{2}}, \quad (4.27)$$

$$\sinh \alpha = \sqrt{\frac{\cosh 2\alpha - 1}{2}}. \quad (4.28)$$

- The sum product identity.
  From (4.22), summing and subtracting the terms with the plus and minus signs and setting $\theta_1 = \alpha + \beta; \theta_2 = \alpha - \beta \Rightarrow \alpha = (\theta_1 + \theta_2)/2; \beta = (\theta_1 - \theta_2)/2$, we obtain

$$\cosh\theta_1 + \cosh\theta_2 = 2\cosh\frac{\theta_1+\theta_2}{2}\cosh\frac{\theta_1-\theta_2}{2},$$

$$\cosh\theta_1 - \cosh\theta_2 = 2\sinh\frac{\theta_1+\theta_2}{2}\sinh\frac{\theta_1-\theta_2}{2}, \qquad (4.29)$$

and from (4.23)

$$\sinh\theta_1 + \sinh\theta_2 = 2\sinh\frac{\theta_1+\theta_2}{2}\cosh\frac{\theta_1-\theta_2}{2},$$

$$\sinh\theta_1 - \sinh\theta_2 = 2\cosh\frac{\theta_1+\theta_2}{2}\sinh\frac{\theta_1-\theta_2}{2}. \qquad (4.30)$$

4.5 Trigonometry in the Minkowski Plane

4.5.1 Analytical Definitions of Hyperbolic Trigonometric Functions

Let us consider the triangle of Fig. 4.3, in the hyperbolic plane, with no sides parallel to axes bisectors, and call $P_n \equiv (x_n, y_n)$; $n = i, j, k$ the vertexes, θ_n the hyperbolic angles.

The quadratic hyperbolic length of the side opposite to vertex P_i is called D_i and defined by (3.5)

$$D_i \equiv D_{jk} = (z_j - z_k)(\tilde{z}_j - \tilde{z}_k)|i \neq j \neq k \quad \text{and} \quad d_i = \sqrt{|D_i|} \qquad (4.31)$$

as pointed out before, D_i must be taken with its sign.

As shown in Fig. 4.3, and following the conventions of Euclidean trigonometry we associate with the sides three vectors oriented from $P_1 \to P_2; P_1 \to P_3; P_2 \to P_3$.

From (4.7) and (4.8), taking into account the sides orientation, we obtain

$$\cosh_e\theta_1 = \frac{(x_2-x_1)(x_3-x_1)-(y_2-y_1)(y_3-y_1)}{d_2 d_3},$$

$$\sinh_e\theta_1 = \frac{(x_2-x_1)(y_3-y_1)-(y_2-y_1)(x_3-x_1)}{d_2 d_3},$$

$$\cosh_e\theta_2 = -\frac{(x_3-x_2)(x_2-x_1)-(y_3-y_2)(y_2-y_1)}{d_1 d_3},$$

$$\sinh_e\theta_2 = \frac{(x_2-x_1)(y_3-y_2)-(y_2-y_1)(x_3-x_2)}{d_1 d_3}, \qquad (4.32)$$

$$\cosh_e\theta_3 = \frac{(x_3-x_2)(x_3-x_1)-(y_3-y_2)(y_3-y_1)}{d_1 d_2},$$

$$\sinh_e\theta_3 = \frac{(x_3-x_1)(y_3-y_2)-(y_3-y_1)(x_3-x_2)}{d_1 d_2}.$$

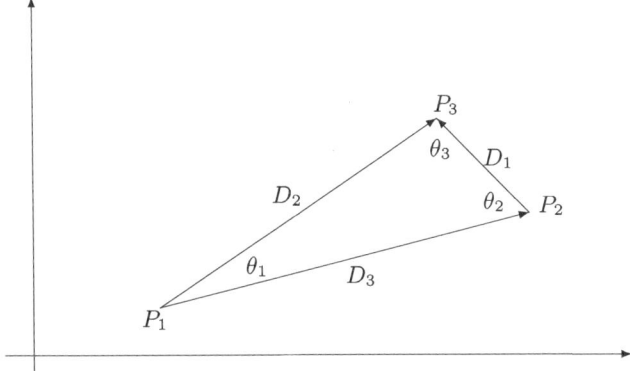

Fig. 4.3 The elements of a triangle in the Cartesian plane. The triangles are representative figures of Euclidean geometry and all the polygons can be considered as composed by triangles. Furthermore the theorems that states the equality of all the elements if just three are equal indicates the *characteristic quantities of Euclidean geometry*: *the lengths of segments* (*the sides*) *and the angles*. If a triangle is represented in a Cartesian plane its characteristic quantities do not change if it is rotated and translated. These operations, called `motions`, can be described by complex numbers (Sect. 2.1). We know that in Euclidean geometry the areas are positive. In analytic geometry we give a sign to the segment length, then the areas can be negative. This could happen in our formalization. In any case the area is positive in this Cartesian representation of a triangle if the sides are oriented as shown in this figure, i.e., associating with the sides three vectors oriented from $P_1 \to P_2; P_1 \to P_3; P_2 \to P_3$. Actually from (4.34) it follows that all the sinus have the same sign then since θ_1 is positive, they are all positive. Let us consider a triangle in the Gauss plane (Fig. 2.1) or in the hyperbolic plane (Fig. 2.3) (in the hyperbolic plane with no sides parallel to axes bisectors). We set $P_n \equiv (x_n, y_n)$; $n = i, j, k$, the vertexes, θ_n the corresponding angles, ρ_n the length of the sides opposite to P_n and v_n the vectors associated with the sides ρ_n. The length of ρ_n (modulus of vector v_n) is given in Euclidean plane by

$$\rho_i = |v_i| = \sqrt{(x_j - x_k)^2 + (y_j - y_k)^2} = \sqrt{(z_j - z_k)(\bar{z}_j - \bar{z}_k)} = |z_j - z_k|, i \neq j \neq k. \qquad \text{The last}$$

expressions, as function of the "complex coordinates" of the vertexes are the same in the complex and hyperbolic planes. From the results shown in Sect. 4.1.1, the trigonometric functions of the angles (4.32), are obtained from the vertexes coordinates by means of (4.2). It follows that the trigonometry theorems, that link the sides with the trigonometric functions of angles, and allow us to obtain all the elements of a triangle from the three ones that determinate it, are just identities. This algebraic description of geometrical theorems allows us a Euclidean formalization of space–time geometry. Actually the introduction of hyperbolic numbers, algebraically equivalent to complex numbers (Fig. 4.1 and Sect. 2.2), but with properties that relate them to Lorentz's group of Special relativity, has allowed us (Sect. 4.2), a complete algebraic formalization of space–time geometry and trigonometry, by removing the lack of intuitive vision of this plane

It is straightforward to verify that all the functions $\sinh_e \theta_n$ have the same numerator. If we call this numerator

$$x_1(y_2 - y_3) + x_2(y_3 - y_1) + x_3(y_1 - y_2) = 2S, \qquad (4.33)$$

we can write

$$2S = d_2d_3 \sinh_e \theta_1 = d_1d_3 \sinh_e \theta_2 = d_1d_2 \sinh_e \theta_3. \tag{4.34}$$

In Euclidean geometry a quantity equivalent to S represents the triangle's area. In hyperbolic geometry, from (4.8) it follows that S is still *an invariant quantity* and from (4.34) that this invariant is *related with the triangle*. For this reason it is appropriate to call S the *pseudo-Euclidean area* [3].

We note that the expression of area (4.33), in terms of vertexes coordinates, is exactly the same as in Euclidean geometry (Gauss' formula for a polygon area applied to a triangle). Therefore the area in the hyperbolic plane can be calculated in a Euclidean way. This coincidence, with reference to Fig. 2.4, its caption and Sect. 2.6.2, allows us to state

Theorem 4.6 *The magnitude of a hyperbolic angle is equal to twice the* hyperbolic or Euclidean area *of the sector OVP of the unit hyperbola.*

4.5.2 Trigonometric Laws in Hyperbolic Plane

Here we see how the theorems of Euclidean trigonometry can be restated for hyperbolic trigonometry.

- Law of sines.

Theorem 4.7 *In a triangle the ratio of the hyperbolic sine to the hyperbolic length of the opposite side is constant*

$$\frac{\sinh_e \theta_1}{d_1} = \frac{\sinh_e \theta_2}{d_2} = \frac{\sinh_e \theta_3}{d_3}. \tag{4.35}$$

Proof This theorem follows from (4.34) if we divide it by $d_1d_2d_3$.   □

As straightforward consequence of (4.35), we have

Theorem 4.8 *If two hyperbolic triangles have the same hyperbolic angles their sides are proportional.*

- Napier's theorem.

Theorem 4.9 *As in Euclidean trigonometry from (4.35), we have*

$$\frac{d_1 + d_2}{d_1 - d_2} = \frac{\tanh_e \frac{\theta_1+\theta_2}{2}}{\tanh_e \frac{\theta_1-\theta_2}{2}}. \tag{4.36}$$

Proof From (4.35), we have

$$\frac{d_1}{d_2} = \frac{\sinh_e \theta_1}{\sinh_e \theta_2} \tag{4.37}$$

from which we can write

$$\frac{d_1 + d_2}{d_1 - d_2} = \frac{\sinh_e \theta_1 + \sinh_e \theta_2}{\sinh_e \theta_1 - \sinh_e \theta_2} \tag{4.38}$$

By applying (4.30) the theorem follows. □

- Carnot's theorem.

Theorem 4.10 *From the definitions of the side lengths* (3.2) *and hyperbolic angular functions given by* (4.32) *we have*

$$D_i = D_j + D_k - 2d_j d_k \cosh_e \theta_i. \tag{4.39}$$

Proof Let us set $i = 1; j = 2; k = 3$. By means of $\cosh_e$ given by (4.32), we can write (4.39) as

$$
\begin{aligned}
(x_3 - x_2)^2 - (y_3 - y_2)^2 &= (x_3 - x_1)^2 - (y_3 - y_1)^2 + (x_2 - x_1)^2 \\
&\quad - (y_2 - y_1)^2 - 2[(x_2 - x_1)(x_3 - x_1) - (y_2 - y_1)(y_3 - y_1)]. \tag{4.40}
\end{aligned}
$$

After reduction of the right-hand side we see that it is the same as the left-hand side. Therefore (4.39) in our formalization is an identity. □

- Law of cosines.

Theorem 4.11 *We have*

$$d_i = |d_j \cosh_e \theta_k + d_k \cosh_e \theta_j|. \tag{4.41}$$

Proof It can be verified as the previous theorem. □

- Pythagoras' theorem.

Theorem 4.12 *For a triangle with the right angle* θ_i, *we have*

$$D_i = D_j + D_k. \tag{4.42}$$

Proof Let us consider a triangle with the side P_iP_k pseudo-orthogonal to P_iP_j and the straight lines given by their extension. By calling m_{ik} and m_{ij} the hyperbolic angular coefficients of the straight lines, we have

$$\frac{y_j - y_i}{x_j - x_i} \equiv m_{ij} = \frac{1}{m_{ik}} \equiv \frac{x_k - x_i}{y_k - y_i},$$

therefore

$$(x_j - x_i)(x_k - x_i) = (y_j - y_i)(y_k - y_i), \tag{4.43}$$

and from (4.32) it follows that $\cosh_e \theta_i = 0 \Rightarrow \theta_i = (0)_h$. Therefore from (4.39) the hyperbolic Pythagoras' theorem (4.42) holds.   □

We note that in the right-hand sides of (4.39) and (4.42) there is a sum of the sides quadratic lengths, as in Euclidean geometry, but in hyperbolic geometry the sides quadratic lengths may be negative. In particular in (4.42) D_j and D_k are pseudo-orthogonal, therefore they have always opposite signs.

- **The Hero's Formula.**
  This formula, in Euclidean geometry, allows us to express the triangle area as a function of the triangle side lengths. Here, taking into account that the characteristic quantities in hyperbolic geometry are the quadratic distances, this formula is expressed as function of them. More precisely, we demonstrate it in an algebraic way for both Euclidean and hyperbolic geometries indicating by u a generic versor that in the following will be specified as i or h. We have

Theorem 4.13 *For the Euclidean and hyperbolic plane geometries the area of a triangle as a function of the sides quadratic lengths can be obtained by the following* generalized Hero's formula

$$S^2 = \frac{(D_1 + D_2 - D_3)^2 - 4D_1D_2}{16u^2} \tag{4.44}$$

$$\equiv \frac{D_1^2 + D_2^2 + D_3^2 - 2(D_1D_2 + D_1D_3 + D_2D_3)}{16u^2} \tag{4.45}$$

Proof We start from the identity

$$(z_1 - z_2)(\bar{z}_1 - \bar{z}_2) \equiv [(z_1 - z_3) - (z_2 - z_3)][(\bar{z}_1 - \bar{z}_3) - (\bar{z}_2 - \bar{z}_3)], \tag{4.46}$$

where $z_i = x_i + u\,y_i$.
Expanding the right-hand side, taking into account (4.31) and, by introducing the quadratic form

$$Q = z_1\bar{z}_2 + z_3\bar{z}_1 + z_2\bar{z}_3 - z_2\bar{z}_1 - z_1\bar{z}_3 - z_3\bar{z}_2, \tag{4.47}$$

we have

$$D_3 = D_1 + D_2 - 2(\bar{z}_2 - \bar{z}_3)(z_1 - z_3) - Q$$
$$\Rightarrow 2(\bar{z}_2 - \bar{z}_3)(z_1 - z_3) = -D_3 + D_1 + D_2 - Q. \tag{4.48}$$

From (4.47), we see that Q has the following property

$$\bar{Q} = -Q, \tag{4.49}$$

and, calculating Q as a function of the points coordinates, we have

$$Q \equiv z_1\bar{z}_2 + z_3\bar{z}_1 + z_2\bar{z}_3 - z_2\bar{z}_1 - z_1\bar{z}_3 - z_3\bar{z}_2$$
$$= (2u)[x_1(y_3 - y_2) + x_2(y_1 - y_3) + x_3(y_2 - y_1)]. \tag{4.50}$$

The content of the square brackets can be recognized, but for the sign, as the double of the area (S) of the triangle, therefore we can write

$$Q^2 = 16u^2 S^2. \tag{4.51}$$

Multiplying (4.48) by its conjugate and taking into account (4.49), we obtain

$$4D_1 D_2 = (-D_3 + D_1 + D_2)^2 - Q^2. \tag{4.52}$$

Substituting in (4.52) the relation (4.51) we obtain the Hero's formula (4.44). This formula can be written in the form (4.45), that is symmetric with respect to the sides quadratic lengths.   □

We note that this demonstration is another example of how the proposed approach allows us to demonstrate theorems by means of identities.

Otherwise we have to point out that while the results of this chapter and Chaps. 5 and 6 are derived from the formalization of hyperbolic trigonometry in Sect. 4.2, this demonstration has required some original steps. In particular the introduction of the identity (4.46), the bilinear expression (4.47) and the relation between Q and the triangle areas (4.51).

Now the results are specified for the two geometries

- For hyperbolic geometry we must set $u^2 \equiv h^2 = 1$.
- For Euclidean geometry the expression (4.44) can be set as a product of four linear terms representing the usual Hero's formula.

Proof By setting $u^2 \equiv i^2 = -1$, calling a, b, c the lengths of the sides of the triangle and $q = (a + b + c)/2$ its semi-perimeter, we write (4.45) as a difference of squared terms and, by means of elementary algebra, we obtain:

$$S^2 = \frac{-a^4 + 2a^2(b^2 + c^2) - (b^2 + c^2)^2 + 4b^2c^2}{16}$$
$$\equiv \frac{-(b^2 + c^2 - a^2)^2 + 4b^2c^2}{16} \equiv q(q-a)(q-b)(q-c) \tag{4.53}$$

□

We have seen that the topology of the hyperbolic plane is more complex with respect to the Euclidean one, as well as the relations between $\sinh_e$ and $\cosh_e$ and between the lengths and the quadratic lengths. For these reasons we could think that the triangle determination would require more information, with respect to the three conditions necessary in Euclidean geometry, nevertheless in Sect. 4.6.3 we shall see that: *All the sides and angles of a hyperbolic triangle can be determined if we know, as in Euclidean geometry, three elements.* We have

Theorem 4.14 *All the elements (sides and angles) of a triangle are invariant for hyperbolic rotation.*

Proof Let us consider a triangle with vertexes in points (see Fig. 4.4)

$$P_1 \equiv (0,0), P_2 \equiv (x_2,0) \quad \text{and} \quad P_3 \equiv (x_3, y_3); \qquad (4.54)$$

since $\overline{P_1 P_3} \equiv d_2$ and $\overline{P_1 P_2} \equiv d_3$ are invariant quantities for hyperbolic rotations, from Theorem 4.3 it follows that θ_1 is invariant too. Since these three elements determine all the other ones, all elements are invariant. □

From this invariance it follows that, by a coordinate axes translation and a hyperbolic rotation, any triangle can be set so that the demonstration of the theorems which follow is facilitated. In particular we set a vertex in $P \equiv (0,0)$ and a side on one coordinate axis, i.e., with vertexes in points given by (4.54 and Fig. 4.4). The sides quadratic lengths are

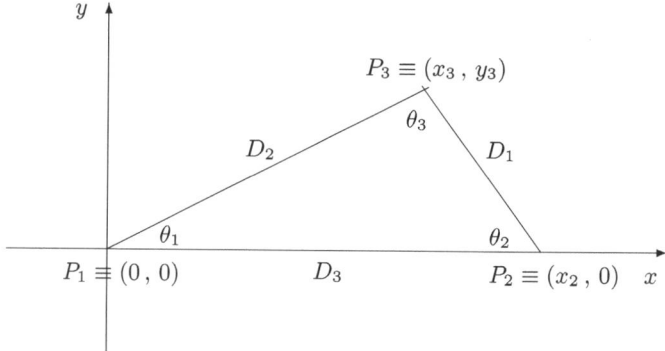

Fig. 4.4 Elements of a triangle in a particular position. The Euclidean geometry is defined (as recalled in Appendix A) as the one that studies the properties of the figures independent of their position in a plane in which the distance between two points is calculated by means of Pythagoras' theorem. In a Cartesian representation this property is equivalent to say that Euclidean geometry studies the properties independent of rotations and translations of the figures or of the coordinate axes. In the same way the space–time geometry is defined. In this geometry the distance is given by a non definite quadratic form and the rotation of Euclidean geometry are replaced by "hyperbolic rotations" (Lorentz transformations of special relativity Sect. 2.4.1). In this geometry, as for Euclidean one (see Theorem 4.14), we can place the Cartesian axes so that the solution of the problems is simplified. In particular we set the axes so that the vertexes of the triangle are in the points (4.54), as shown in this figure

$$D_1 = (x_3 - x_2)^2 - y_3^2; \quad D_2 = x_3^2 - y_3^2; \quad D_3 = x_2^2. \tag{4.55}$$

By using (4.31) and (4.32), we obtain the other elements:

$$\cosh_e \theta_1 = \frac{x_2 x_3}{d_2 d_3}; \quad \cosh_e \theta_2 = \frac{x_2 (x_2 - x_3)}{d_1 d_3}; \quad \cosh_e \theta_3 = \frac{x_3 (x_3 - x_2) - y_3^2}{d_1 d_2};$$

$$\sinh_e \theta_1 = \frac{x_2 y_3}{d_2 d_3}; \quad \sinh_e \theta_2 = \frac{x_2 y_3}{d_1 d_3}; \quad \sinh_e \theta_3 = \frac{x_2 y_3}{d_1 d_2}. \tag{4.56}$$

- The Triangle's Angles Sum
  In a Euclidean triangle, given two angles (ϕ_1, ϕ_2), the third one (ϕ_3) can be found using the relation

$$\phi_1 + \phi_2 + \phi_3 = \pi. \tag{4.57}$$

Since π is closely related with the circle in Euclidean geometry, we could not assume that a similar mathematical expression holds in hyperbolic geometry, but now we see that the hyperbolic angles of a triangle are linked by a relation that can be considered as equivalent to (4.57).

Actually (4.57) can be expressed in the form

$$\sin(\phi_1 + \phi_2 + \phi_3) = 0, \quad \cos(\phi_1 + \phi_2 + \phi_3) = -1$$

that allows us to verify

Theorem 4.15 *By means of the formalism exposed in Sect. 4.3.1, summarized by (4.19), we can state: the sum of the triangle's angles is given by*

$$(\theta_1)_k + (\theta_2)_{k'} + (\theta_3)_{k''} \equiv (\theta_1 + \theta_2 + \theta_3)_{k \cdot k' \cdot k''} = (0)_{\pm 1}. \tag{4.58}$$

Proof Exploiting (4.21) and using (4.56), we obtain

$$\sinh_e(\theta_1 + \theta_2 + \theta_3) \equiv \sinh_e \theta_1 \sinh_e \theta_2 \sinh_e \theta_3 + \sinh_e \theta_1 \cosh_e \theta_2 \cosh_e \theta_3$$

$$+ \cosh_e \theta_1 \sinh_e \theta_2 \cosh_e \theta_3 + \cosh_e \theta_1 \cosh_e \theta_2 \sinh_e \theta_3$$

$$\equiv \frac{x_2^2 y_3 [x_2 y_3^2 + x_2 x_3 (x_2 - x_3) + x_3^2 (x_3 - x_2)]}{d_1^2 d_2^2 d_3^2}$$

$$- \frac{x_2^2 y_3 [x_3 y_3^2 - x_3 (x_2 - x_3)^2 - y_3^2 (x_2 - x_3)]}{d_1^2 d_2^2 d_3^2} = 0,$$

$$\cosh_e(\theta_1 + \theta_2 + \theta_3) \equiv \cosh_e\theta_1 \cosh_e\theta_2 \cosh_e\theta_3 + \sinh_e\theta_1 \sinh_e\theta_2 \cosh_e\theta_3$$
$$+ \sinh_e\theta_1 \cosh_e\theta_2 \sinh_e\theta_3 + \cosh_e\theta_1 \sinh_e\theta_2 \sinh_e\theta_3$$
$$\equiv \frac{x_2^2\{-x_3^2(x_2-x_3)^2 + y_3^2[-x_3(x_2-x_3)+x_2x_3]\}}{d_1^2 d_2^2 d_3^2}$$
$$+ \frac{y_3^2[x_2(x_2-x_3)-x_3(x_2-x_3)-y_3^2]}{d_1^2 d_2^2 d_3^2}$$
$$\equiv \frac{x_2^2\{-x_3^2[(x_2-x_3)^2-y_3^2]+y_3^2[(x_2-x_3)^2-y_3^2]\}}{d_1^2 d_2^2 d_3^2}$$
$$\equiv -\frac{D_1 D_2 D_3}{d_1^2 d_2^2 d_3^2} = \pm 1. \tag{4.59}$$

Therefore

$$\sinh_e(\theta_1 + \theta_2 + \theta_3) = 0, \quad \cosh_e(\theta_1 + \theta_2 + \theta_3) = \pm 1, \tag{4.60}$$

that are equivalent to (4.58). □

This result allows us to state that if we know two angles we can determine if the Klein's group index of the third angle is $\pm h \to \{\theta \in Us, Ds\}$ or $\pm 1 \to \{\theta \in Rs, Ls\}$, and obtain the relation between $\cosh_e$ and $\sinh_e$ as stated by (4.15). This relation and the condition (4.58) allow us to obtain the hyperbolic functions of the third angle. Therefore we have

Theorem 4.16 *Also for hyperbolic triangles, if we know two angles we can obtain the third one (see example 2 in Sect. 4.6.3).*

In the following section we show some examples of solutions for general hyperbolic triangles.

4.6 Triangles in Hyperbolic Plane

4.6.1 The Triangular Property

In the hyperbolic geometry the triangular property of Euclidean geometry is reversed and the following theorem holds

Theorem 4.17 *In a triangle with sides of the first kind, the largest side is larger than the sum of the other two sides.*

Proof This inequality follows at once from the law of cosines (4.41). Actually for sides of the first kind we can delete the absolute value and to substitute cosh to $\cosh_e$. Therefore we have

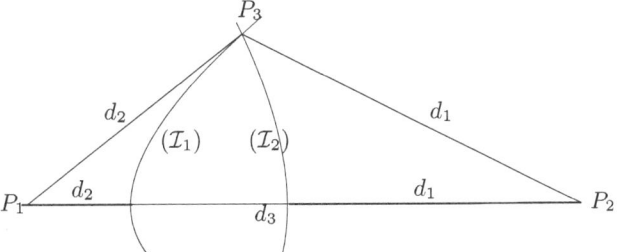

Fig. 4.5 Geometrical representation of triangular inequality in the hyperbolic plane. In Euclidean geometry the lengths of two segments with a common point can be compared by reporting one on the other by means of a circle with center in their common point, as well as in hyperbolic geometry two segment of the same kind can be compared by means of an equilateral hyperbola. In particular for the triangle represented in this figure, the side d_2 is reported on the side d_3 by means of the equilateral hyperbola ($\mathcal{I}_1$) with center in P_1 and semidiameter d_2. In a similar way the side d_1 is reported on the side d_3 by means of the equilateral hyperbola ($\mathcal{I}_2$) with center in P_2 and semidiameter d_1. We can check at once that $d_3 > d_1 + d_2$

$$d_3 = d_2 \cosh \theta_1 + d_1 \cosh \theta_2. \tag{4.61}$$

Since $\cosh \theta_i > 1$, we have $d_3 > d_2 + d_1$.

This property can also be inspected by means of an "Euclidean-like" geometrical construction (Fig. 4.5).

Actually we observe that for comparing lengths in hyperbolic geometry the sides d_2 and d_1 can be reported on the side d_3 by means of equilateral hyperbolas with center in P_1 and semidiameter d_2 and center in P_2 and semidiameter d_1, respectively.

Therefore this reverse triangular inequality derives from the substitution, in hyperbolic geometry, of equilateral hyperbolas to Euclidean circle. $\square$

This property shall be seen again (Sect. 6.4) by considering also curved lines.

4.6.2 The Elements in a Right-Angled Triangle

Let us consider a triangle as shown in Fig. 4.4 with a right-angle in the vertex P_2.

This particular triangle allows us to explain some of the concepts came to light in formalizing the hyperbolic trigonometry. In particular

- The meaning of Klein's index, introduced in Sect. 4.3.1. By applying it to Euclidean plane we obtain some known results.
- The sum of hyperbolic angles in a triangle.

4.6.2.1 Application of Klein' Index to the Sum of Triangle Angles

Let us consider a right-angled triangle in the position represented in Fig. 4.4

- Euclidean geometry
  Let us call α_i the angles in the vertexes P_i, and define them as "extended angles" (see Sect. 4.3.1). If we indicate with a subscript the "Euclidean Klein's group index" $(k \in K = \{1, i, -1, -i\})$, we have $\alpha_1 = (\alpha_1)_1, \alpha_2 \equiv \pi/2 = (0)_i, \alpha_3 \equiv \pi/2 - \alpha_1 = (-\alpha_1)_i$.   Therefore $\alpha_1 + \alpha_2 + \alpha_3 \equiv (\alpha_1)_1 + (0)_i + (-\alpha_1)_i \equiv (\alpha_1 + 0 - \alpha_1)_{i \cdot i} = (0)_{-1} \equiv \pi$, as it is well known.
- Hyperbolic geometry
  Now we consider the same triangle in hyperbolic plane. Setting in (4.56) $x_2 = x_3$, we have

$$
\begin{cases}
D_1 = -y_3^2 \\
\cosh_e \theta_1 = x_3/d_2 \\
\sinh_e \theta_1 = y_3/d_2
\end{cases}
\qquad
\begin{cases}
D_2 = x_3^2 - y_3^2 \\
\cosh_e \theta_2 = 0 \\
\sinh_e \theta_2 = 1
\end{cases}
\qquad
\begin{cases}
D_3 = x_3^2 \\
\cosh_e \theta_3 = -y_3/d_2 \\
\sinh_e \theta_3 = x_3/d_2
\end{cases}
$$

It results that $\theta_2 = (0)_h$ and we consider the following cases:

1. $P_3 \in Rs \Rightarrow \theta_1 = (\theta_1)_1$ and by means of Table 4.1 we have $\theta_3 = (-\theta_1)_h$. Therefore $\theta_1 + \theta_2 + \theta_3 \equiv (\theta_1)_1 + (0)_h + (-\theta_1)_h \equiv (\theta_1 + 0 - \theta_1)_{h \cdot h} = (0)_1$.
2. $P_3 \in Us$. Setting $\theta_3 = -\theta'$ we have $\cosh_e \theta' = -y_3/d_2, \sin \theta' = -x_3/d_2$. Therefore $\theta' = (\theta')_1 \Rightarrow \theta_3 = (-\theta')_{-1}$, and from Table 4.1 we have $\theta_1 = (\theta')_h$, and $\theta_1 + \theta_2 + \theta_3 \equiv (\theta')_h + (0)_h + (-\theta')_{-1} = (0)_{-1}$.

We have found the two possibilities of the Klein index shown by Theorem 4.15 in Sect. 4.5.2 for the sum of the triangle angles.

4.6.3 Solution of Hyperbolic Triangles

In this section, in order to point out analogies and differences with Euclidean trigonometry, we report some examples in which we determine the elements of hyperbolic triangles. We shall note that the Cartesian representation can give some simplifications in the triangle solution.

The Cartesian axes will be chosen as shown in Fig. 4.4: $P_1 \equiv (0, 0)$, $P_2 \equiv (\pm d_3, 0)$, or $P_2 \equiv (0, \pm d_3)$, where $\overline{P_1 P_2} = d_3$ is the side (or one of the sides) which we know. The two possibilities for P_2 depend on the D_3 sign, the sign plus or minus is chosen so that one goes from P_1 to P_2 to P_3 anticlockwise. Thanks to relations (4.55) and (4.56), the triangle is completely determined by the coordinates of point P_3, therefore the solution is obtained by finding these coordinates.

1. Given two Sides and the Angle Between Them

   Let be given θ_1; D_2; D_3.

   For D_3 positive we put P_2 on the x axis, therefore $P_2 \equiv (\pm d_3, 0)$, the sign of d_3 being such that P_1, P_2, P_3 follow each other anticlockwise. In this way $\sinh \theta_i$ and the triangle's area are positive, as in Euclidean trigonometry in a Cartesian representation. So we have

   $$x_3 = d_2 \cosh_e \theta_1; \quad y_3 = d_2 \sinh_e \theta_1, \tag{4.62}$$

   and (4.56) allow us to determine the other elements.

   For D_3 negative, we must take P_2 on the y axis and in (4.62) we must change $\sinh_e \theta_1 \leftrightarrow \cosh_e \theta_1$. Grouping together both examples, we have

   $$\begin{aligned} \text{if } D_3 > 0 & \quad x_3 = d_2 \cosh_e \theta_1; \quad y_3 = d_2 \sinh_e \theta_1; \\ \text{if } D_3 < 0 & \quad x_3 = d_2 \sinh_e \theta_1; \quad y_3 = d_2 \cosh_e \theta_1. \end{aligned} \tag{4.63}$$

2. Given two Angles and the Side Between Them

   Let be given θ_1; θ_2; D_3.

   In Euclidean trigonometry the solution of this problem is obtained by using the condition that the sum of the three angles is π. We use this method which allows us to use the Klein group defined in Sect. 4.3.1, as well as a method peculiar to analytical geometry.

   Let us start with the classical method: the Klein indexes (k_1, k_2) of known angles θ_1, θ_2 are obtained by means of (4.20). For the third angle we have $\theta_3 = -(\theta_1 + \theta_2)$ and k_3 is such that $k_1 k_2 k_3 = \pm 1$. Therefore the hyperbolic trigonometric functions are given by $\sinh_e \theta_3 = \sinh|\theta_1 + \theta_2|$ if $k_3 = \pm 1$ and by $\sinh_e \theta_3 = \cosh(\theta_1 + \theta_2)$ if $k_3 = \pm h$. Now we can apply the law of sines and obtain $d_2 = d_3 \frac{\sinh_e \theta_2}{\sinh_e \theta_3}$. Equations 4.62 allow us to obtain the P_3 coordinates.

   In the Cartesian representation we can use the following method: let us consider the straight lines $P_1 P_3$ and $P_2 P_3$. Solving the algebraic system between these straight lines we obtain the P_3 coordinates.

   For $D_3 > 0$ the straight line equations are $P_1 P_3 \Rightarrow y = x \tanh_e \theta_1$ and $P_2 P_3 \Rightarrow y = -(x - x_2) \tanh_e \theta_2$, and we obtain

   $$P_3 \equiv \left(x_2 \frac{\tanh_e \theta_2}{\tanh_e \theta_2 + \tanh_e \theta_1}, x_2 \frac{\tanh_e \theta_1 \tanh_e \theta_2}{\tanh_e \theta_2 + \tanh_e \theta_1} \right). \tag{4.64}$$

   For $D_3 < 0$ the straight line equations are $y = x \coth_e \theta_1$ and $y - y_2 = -x \coth_e \theta_2$, with the solution

   $$P_3 \equiv \left(y_2 \frac{\tanh_e \theta_1 \tanh_e \theta_2}{\tanh_e \theta_2 + \tanh_e \theta_1}, y_2 \frac{\tanh_e \theta_2}{\tanh_e \theta_2 + \tanh_e \theta_1} \right). \tag{4.65}$$

3. Given two Sides and one Opposite Angle

   Let be given θ_1; D_1; D_3, with $D_3 > 0$.

   We know, that in Euclidean geometry, these elements do not determine unequivocally a triangle. The same happens in hyperbolic geometry.

Applying the Carnot's theorem (4.39) to the side d_1 we have $D_2 - 2d_2d_3 \cosh_e \theta_1 + D_3 - D_1 = 0$, from which we can obtain d_2. Actually for $\cosh_e \theta_1 > \sinh_e \theta_1$

$$D_2 = d_2^2 \Rightarrow d_2 = d_3 \cosh_e \theta_1 \pm \sqrt{d_3^2 \sinh_e^2 \theta_1 + D_1},$$

for $\cosh_e \theta_1 < \sinh_e \theta_1$

$$D_2 = -d_2^2 \Rightarrow d_2 = -d_3 \cosh_e \theta_1 \pm \sqrt{d_3^2 \sinh_e^2 \theta_1 - D_1}. \tag{4.66}$$

So, as for the equivalent Euclidean problem, we can have, depending on the value of the square root argument, two, one or no solutions. The coordinates of the vertex P_3 are given by (4.62).

Now we use an analytic method typical of the Cartesian plane. The coordinates of P_3 can be obtained by intersecting the straight line $y = x \tanh_e \theta_1$ with the hyperbola centered in P_2 and having quadratic semi-diameter $P = D_1$, i.e., by solving the system

$$y = x \tanh_e \theta_1; \quad (x - d_3)^2 - y^2 = D_1. \tag{4.67}$$

The results are the same of (4.66), but now it is easy to understand the geometrical meaning of the solutions which can be compared with the equivalent Euclidean problem with an equilateral hyperbola instead of a circle.

Actually if $D_1 > 0$ and $d_1 > d_3$ the point P_1 is included in a hyperbola arm and we have always two solutions. Otherwise, if $\sinh \theta_1 < d_1/d_3$ there are no solutions, if $\sinh \theta_1 = d_1/d_3$ there is just one solution and if $\sinh \theta_1 > d_1/d_3$ two solutions.

If $D_3 < 0$ the P_2 vertex must be put on the y axis and we have the system

$$x = y \tanh_e \theta_1; \quad (y - d_3)^2 - x^2 = -D_1.$$

Comparing this result with the solutions of system (4.67) we have to change $x \leftrightarrow y$.

4. Given two Angles and one Opposite Side
   Let be given θ_1; θ_3; D_3 with $D_3 > 0$.
   From theorems (4.35) and (4.41), we have

$$d_1 = d_3 \frac{\sinh_e \theta_1}{\sinh_e \theta_3}; \quad |d_2| = |d_1 \cosh_e \theta_3 + d_3 \cosh_e \theta_1|. \tag{4.68}$$

From (4.62) we find P_3 coordinates.

5. Given three Sides
   From the Carnot's law of cosine (4.39), we have

$$\cosh_e \theta_1 = \frac{D_2 + D_3 - D_1}{2d_2d_3}. \tag{4.69}$$

For $D_3 > 0$, we take $P_1 \equiv (0,0), P_2 \equiv (\pm\sqrt{D_3}, 0)$ and $\sinh_e \theta_1$ is given by

- if $D_2 > 0 \Rightarrow \sinh_e \theta_1 = \sqrt{\cosh_e^2 \theta_1 - 1}$,
- if $D_2 < 0 \Rightarrow \sinh_e \theta_1 = \sqrt{\cosh_e^2 \theta_1 + 1}$.

From (4.62) we find P_3 coordinates.

References

1. F. Catoni, R. Cannata, V. Catoni, P. Zampetti, Hyperbolic trigonometry in two-dimensional space-time Geometry. Nuovo Cimento. B. **118**(5), 475 (2003)
2. P. Fjelstad, Extending relativity via the perplex numbers, Am. J. Phys. **54**, 416 (1986)
3. I.M. Yaglom, *A Simple Non-Euclidean Geometry and its Physical Basis* (Springer, New York, 1979)

Chapter 5
Equilateral Hyperbolas and Triangles in the Hyperbolic Plane

Abstract The equilateral hyperbolas, represented in the Minkowski space-time, hold the same properties of circles in Euclidean plane and satisfy similar theorems. At the same time equivalent relations to the ones in Euclidean plane between circles and triangles are obtained in hyperbolic plane between equilateral hyperbolas and triangles.

Keywords Geometry in hyperbolic plane · Equilateral hyperbola in space-time · Hyperbolas and triangles in space-time

5.1 Theorems on Equilateral Hyperbolas

We have seen in Chap. 3 that, in the hyperbolic plane, the four arms of the unit equilateral hyperbolas $x^2 - y^2 = \pm 1$ correspond to the unit circle, for the definition of trigonometric functions. Indeed the equilateral hyperbolas have many of the properties of circles in the Euclidean plane; here we point out some of them, showing some theorems in the hyperbolic plane that are the equivalent ones of well-known theorems holding for the circle in the Euclidean plane.

Definitions If A and B are two points lying on an equilateral hyperbola, the segment AB is called a *chord* of the hyperbola. We define two kinds of chords: if points A, B are on the same arm of the hyperbola we have "*external chords*", if the points are in opposite arms we have "*internal chords*".

We extend these definitions to points: given a two-arms hyperbola we call *external* the points inside the arms, *internal* the points between the arms and the axes bisectors.

This last definition agrees with the one for circles in which the center is an internal point and the distance of these points from the center is lesser than the radius.

Any internal chord which passes through the center "P_c" of the hyperbola is called a *diameter of the hyperbola*. We call p the semi-diameter and P the "quadratic semi-diameter", with its sign ($p = \sqrt{|P|}$).

Here we extend to equilateral hyperbolas the definitions stated for segments and straight lines and call them *hyperbolas of the first (second) kind if the tangent straight lines are of the first (second) kind*. Actually, as far as a general curve in hyperbolic plane is concerned, (for example a circle) we can not assign it a kind since general curves may have tangent straight lines of both kinds. Only equilateral hyperbolas have the peculiar property that the tangent straight lines to a given arm are of the same kind. This allows us to attribute a kind, depending on the P sign, to the hyperbolas arms. Therefore we have hyperbolas of the first (second) kind if $P < 0$ ($P > 0$), respectively.

The parametric equations of a general equilateral hyperbola are the same as those for the circles with the usual substitutions of extended hyperbolic to circular trigonometric functions. So they are given by

$$x = x_c \pm p \cosh_e \theta, \quad y = y_c \pm p \sinh_e \theta \qquad (5.1)$$

and depend on three parameters: the center's coordinates $P_c \equiv (x_c, y_c)$ and the half-diameter p. This hyperbola is determined by three conditions as the equations for circles. In particular these three conditions can be the passage through three non-aligned points.

Now we enunciate for equilateral hyperbolas the hyperbolic counterpart of the well-known Euclidean theorems for circles. The demonstration of these theorems is performed by elementary analytic geometry.

Theorem 5.1 *The axis of two points on an equilateral hyperbola passes through the center (x_c, y_c).*

An equivalent form is: *The line joining P_c with the midpoint M of a chord is pseudo-orthogonal to it.*

Proof Let us consider two points P_1, P_2 on the same arm of equilateral hyperbola (5.1), determined by hyperbolic angles θ_1, θ_2, and calculate the axis of $\overline{P_1 P_2}$. By substituting in (3.16) the coordinates given by (5.1), using (4.29) and (4.30) in the third passage, we obtain

$$(y - y_c) = (x - x_c) \frac{\cosh_e \theta_1 - \cosh_e \theta_2}{\sinh_e \theta_1 - \sinh_e \theta_2}$$
$$\equiv (x - x_c) \frac{2\sinh_e \frac{\theta_1 - \theta_2}{2} \sinh_e \frac{\theta_1 + \theta_2}{2}}{2\sinh_e \frac{\theta_1 - \theta_2}{2} \cosh_e \frac{\theta_1 + \theta_2}{2}} \equiv (x - x_c) \tanh_e \frac{\theta_1 + \theta_2}{2}. \qquad (5.2)$$

Equation 5.2 demonstrates the theorem.   □

In a similar way we can find that if the points are on different arms of the hyperbola we have just to change (5.2) with $(y - y_c) = (x - x_c) \coth_e [(\theta_1 + \theta_2)/2]$.

This theorem also holds in the limiting position when the points are coincident and the chord becomes tangent to the hyperbola, and we have

Theorem 5.2 *For points M on equilateral hyperbola, the tangent at M is pseudo-orthogonal to the diameter P_cM.*

For the demonstration of the following theorems, we do not lose in generality if we consider a hyperbola of the second kind, with its center at the coordinate origin. We also set $P_c \to O$ and (5.1) becomes

$$x = \pm p \cosh \theta; \quad y = \pm p \sinh \theta. \tag{5.3}$$

Theorem 5.3 *The diameters of a hyperbola are the internal chord of minimum length.*

Proof The "internal chords" joint two points on different arms of the hyperbola. For these points we have

$$A \equiv (p \cosh \theta_1, \ p \sinh \theta_1), \quad B \equiv (-p \cosh \theta_2, -p \sinh \theta_2).$$

By applying (4.22) and (4.27) in the last passage, the quadratic length of the chord is

$$\overline{AB}^2 = p^2[(\cosh \theta_1 + \cosh \theta_2)^2 - (\sinh \theta_1 + \sinh \theta_2)^2] \equiv 4p^2 \cosh^2[(\theta_1 - \theta_2)/2],$$

therefore

$$\overline{AB}^2 = 4p^2 \quad \text{for} \ \theta_1 = \theta_2; \qquad \overline{AB}^2 > 4p^2 \quad \text{for} \ \theta_1 \neq \theta_2. \qquad \square$$

Theorem 5.4 *If points A and B lie on the same arm of a hyperbola, for any point P between A and B, the hyperbolic angle $\widehat{APB}$ is half the hyperbolic angle $\widehat{AOB}$.*

Proof We begin by considering all points on the arm of hyperbola $\in Rs$.

Referring to Fig. 5.1 let us take points $A \equiv (p \cosh \theta_A, \ p \sinh \theta_A)$, $B \equiv (p \cosh \theta_B, \ p \sinh \theta_B)$ with $\theta_A < \theta_B$ and the hyperbola arc between them.

If $P \equiv (p \cosh \theta, \ p \sinh \theta)$ is a point on this arc, i.e., $\theta_A < \theta < \theta_B$, we call α the hyperbolic angle $\widehat{APB}$. The points $A \to P \to B$ follow each other in clockwise direction, therefore they correspond to points P_2, P_1, P_3 of Fig. 4.3, respectively. Therefore from (4.32) we have

$$
\begin{aligned}
\tanh \alpha &\equiv \frac{\sinh \alpha}{\cosh \alpha} \\
&= \frac{(\cosh \theta_B - \cosh \theta)(\sinh \theta_A - \sinh \theta) - (\cosh \theta_A - \cosh \theta)(\sinh \theta_B - \sinh \theta)}{(\cosh \theta_A - \cosh \theta)(\cosh \theta_B - \cosh \theta) - (\sinh \theta_A - \sinh \theta)(\sinh \theta_B - \sinh \theta)}.
\end{aligned}
\tag{5.4}
$$

The differences between trigonometric functions in round brackets can be written, by means of (4.29) and (4.30), as products. After reduction, we obtain

$$\tanh \alpha = \tanh \frac{\theta_B - \theta_A}{2}, \tag{5.5}$$

independent of θ, i.e., of the point on the hyperbola arc.

Now we call β the angle $\widehat{AOB}$. Points $A \to O \to B$ follow each other in clockwise direction and, referring to Fig. 4.3, from (4.32) we have

$$\tanh \beta = -\frac{\cosh \theta_A \sinh \theta_B - \cosh \theta_B \sinh \theta_A}{\cosh \theta_A \cosh \theta_B - \sinh \theta_A \sinh \theta_B} = \tanh(\theta_B - \theta_A). \tag{5.6}$$

From (5.5) and (5.6) the theorem follows.   □

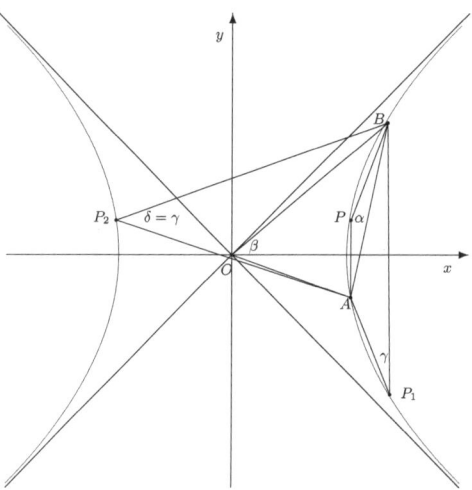

Fig. 5.1 Angles that subtend the same chord of an equilateral hyperbolas. By means of Theorems 5.4 and 5.5 it has been shown that for the angles α relative to points P on the hyperbola arc that subtends the chord $\overline{AB}$, we have in hyperbolic plane, similar relations to the ones holding in Euclidean plane. • are the half of central angle β, • are complementary, i.e., are equal, but with opposite sign of Klein index, to angles (γ) determinated by points P_1 external to chord, • among these last ones are also included points P_2 on the arm $\in Ls$ of hyperbolas. This result is the same of projective geometry. Actually the hyperbola arms $\in Ls$ and $\in Rs$, connect each other in the points at ∞ (see Fig. 3.1). • the relation between the angles α and γ also holds if P and $P_1 \to A$. Now we observe some equivalence with Euclidean geometry. (1) Let us consider the straight line determined by the points A and B: the angles on the hyperbola that are half the central angle are determined by points P on the arc in the same side of the center with respect to straight line $A - B$. (2) The relation between the angles α and γ can be seen in a "Euclidean way" by considering the points $P \to A$ or $P_1 \to A$, i.e., from the internal or external points of the chord. Referring to Theorem 5.5, we have the angle between the tangent line (τ) and the chord (σ) is α if $P \to A$ from the internal points, while by considering $P_1 \to A$ from the external points, the angle between τ and σ is γ supplementary of α

Theorem 5.5 *A complete equivalence of Theorem 5.4, with the analogous Euclidean theorem can also be stated by considering the supplementary angles and the chords becoming tangent.*

Proof

1. Referring to Fig. 5.1, let us consider a point P_1 outside the arc AB. Points $A \rightarrow P_1 \rightarrow B$ follow each other in anticlockwise direction therefore they correspond to points P_2, P_1, P_3 of Fig. 4.3, respectively, and calling γ the angle $\widehat{APB}$, from (4.32) we obtain

$$\tanh \gamma = \tanh \frac{\theta_A - \theta_B}{2} = -\tanh \frac{\theta_B - \theta_A}{2} \Rightarrow \alpha = -\gamma.$$

   In the language of Klein's index (Sect. 4.3.1), if $\tanh \alpha = -\tanh \gamma$ we have $(\alpha)_k = (\gamma)_{-k}$, therefore $\alpha + \gamma = (\alpha + \gamma)_{-k \cdot k} \equiv (0)_{-1}$. This relation corresponds to the Euclidean case for which $\alpha + \gamma = \pi$.

2. Let us consider a point $P_2 \equiv (-p \cosh \theta, -p \sinh \theta)$ on the hyperbola arm $\in Ls$. By calling δ the angle $\widehat{AP_2B}$, with similar calculations of (5.4), we obtain $\tanh \delta = \tanh(\theta_A - \theta_B)/2 \Rightarrow \delta = \gamma$.

3. We complete the parallelism with Euclidean geometry by taking $P \equiv A$. This case allows us to see in "a Euclidean way" the relation between α and γ. Actually let us consider the point $P \rightarrow A$ from the internal points of the chord, the angle between the tangent line (τ) and the chord (σ) is α. Otherwise if $P_1 \rightarrow A$ (from external points), the angle between τ and σ is γ, i.e., the supplementary angle of α. We demonstrate the first possibility. We have

Fig. 5.2 Right-angle triangle inscribed in a hyperbola. In hyperbolic geometry the following Theorem (5.7) holds: If a side of a triangle inscribed in an equilateral hyperbola passes through the center of the hyperbola, the other two sides are pseudo-orthogonal. In this figure the sides $\overline{P_1 P_2}$ and $\overline{P_2 P_3}$ are pseudo-orthogonal

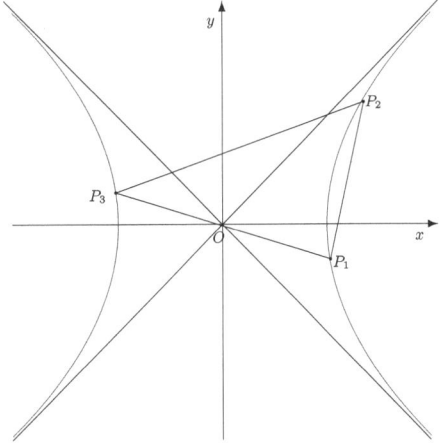

$$\text{angular coefficient of } \tau \Rightarrow \coth \theta_A \equiv \tanh \mu, \tag{5.7}$$

$$\text{angular coefficient of } \sigma \Rightarrow \frac{\sinh \theta_A - \sinh \theta_B}{\cosh \theta_A - \cosh \theta_B} \equiv \tanh v, \tag{5.8}$$

we obtain

$$\tanh \alpha \equiv \frac{\tanh \mu - \tanh v}{1 - \tanh \mu \tanh v} = \frac{1 - \cosh(\theta_A - \theta_B)}{\sinh(\theta_A - \theta_B)} \equiv -\tanh \frac{\theta_A - \theta_B}{2}, \tag{5.9}$$

which completes the assertion.   □

As a straightforward consequence of Theorem 5.4 and, in particular from relation (5.5), we have

Theorem 5.6 *If Q is a second point on the hyperbola between A and B, we have* $\widehat{APB} = \widehat{AQB}$.

Theorem 5.7 *If a side of a triangle inscribed in an equilateral hyperbola passes through the center of the hyperbola, the other two sides are pseudo-orthogonal (Fig. 5.2).*

Proof For two points on an equilateral hyperbola and on a line passing through the center, the coordinates are given by

$$P_1 \equiv (p \cosh \theta_1, p \sinh \theta_1), \quad P_3 \equiv (-p \cosh \theta_1, -p \sinh \theta_1).$$

Let us consider a third point on the hyperbola $P_2 \equiv (p \cosh \theta_2, p \sinh \theta_2)$ and the straight lines determined by the sides $P_1 P_3$, $P_2 P_3$. The equations of these straight lines are given by

$$P_1 P_3 \Rightarrow \frac{y - p \sinh \theta_1}{\sinh \theta_2 - \sinh \theta_1} = \frac{x - p \cosh \theta_1}{\cosh \theta_2 - \cosh \theta_1},$$

$$P_2 P_3 \Rightarrow \frac{y + p \sinh \theta_1}{\sinh \theta_2 + \sinh \theta_1} = \frac{x + p \cosh \theta_1}{\cosh \theta_2 + \cosh \theta_1}.$$

The product of the angular coefficients of these straight lines is 1, therefore they are pseudo-orthogonal.   □

Theorem 5.8 *If from a non-external point $P \equiv (x_p, y_p)$ we trace a tangent and a secant line to a hyperbola, we have: the square of the distance of the tangent point is equal to the product of the distances of secant points.*

A "Euclidean demonstration" of this theorem is reported in the captions of Figs. 5.3 and 5.4, here we see a demonstration by means of the analytic method exposed in this chapter. In this way it is automatically extended to secant lines crossing the different arms of an equilateral hyperbola.

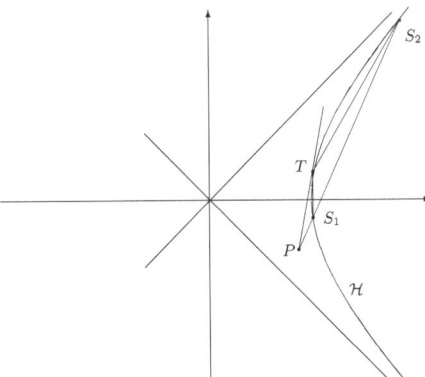

Fig. 5.3 Tangent and secant lines to an arm of equilateral hyperbola (Theorem 5.8). In this and next figure we represent in the hyperbolic plane the corresponding theorem of Euclidean geometry concerning tangent and secant lines to a circle. This theorem can be demonstrated either in analytical or in the same geometrical way of Euclidean plane. In this caption we show this last method, in the text we see the other one. Let us consider the triangles PS_2T and PTS_1; they are similar since: • The angles $\widehat{S_2PT}$ and $\widehat{S_1PT}$ are in common. • The angles $\widehat{PTS_1}$ and $\widehat{TS_2S_1}$ subtend the same hyperbola arc $\widehat{TS_1}$; therefore from Theorem 5.4, are equal. • Third angles are equal for the relations (4.60). Therefore from Theorem 4.8 the triangles PTS_1 and PS_2T have proportional sides, and as for Euclidean geometry, it follows: $\overline{PS_1} \cdot \overline{PS_2} = \overline{PT}^2$. It also follows that the product $\overline{PS_1} \cdot \overline{PS_2}$ is independent of straight lines passing through point P

Proof Let us set

$$\overline{PT} \equiv t; \qquad \overline{PS_l} \equiv s_l; \quad l = 1, 2$$

and consider the hyperbola

$$x^2 - y^2 = 1, \tag{5.10}$$

the non-external point $P \equiv (x_p, y_p)$, and the straight line

$$y - y_p = m(x - x_p) \equiv (x - x_p)\tanh_e \theta_p. \tag{5.11}$$

Eliminating y between (5.10) and (5.11) we obtain, for the abscissas of the intersection points, the equation

$$(1 - m^2)x^2 + 2\,m\,x(m\,x_p + y_p) - m^2x_p^2 - 2\,m\,x_p\,y_p - y_p^2 - 1 = 0. \tag{5.12}$$

Now let us see that it is not necessary to find the roots of this equation. Actually setting $S_l \equiv (x_l, y_l)$, $l = 1, 2$, we have

$$s_l = \frac{|x_l - x_p|}{|\cosh_e \theta_p|} \equiv |(x_l - x_p)|\sqrt{|1 - m^2|} \tag{5.13}$$

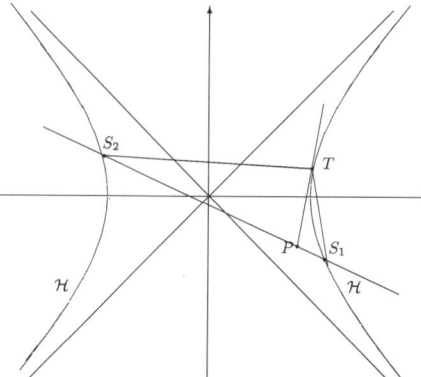

Fig. 5.4 Secant line on the different arms of the same kind of an equilateral hyperbola. In the text (Theorem 5.9) is demonstrated in a "Euclidean" way that the product $\overline{PS_1} \cdot \overline{PS_2}$ is independent of the line passing through P and intersecting the hyperbola arms of the same kind, here we report the "Euclidean-like" demonstration for the secant line crossing different arms of the hyperbola. • The angles $\widehat{S_2PT}$ and $\widehat{S_1PT}$, seen in a Euclidean way are supplementary angles. In the hyperbolic plane, if we call k the Klein index of the first angle, the index of the second one is $-k$ with the same value of the angle. • The angles $\widehat{PTS_1}$ and $\widehat{PS_2T}$ subtend the same hyperbola arc $\overset{\frown}{TS_1}$; therefore, from Theorem 5.4, point 2, they are equal except for the Klein index. Actually by calling k_1 the Klein index of the first angle, the index of the second one is $-k_1$. • The third angles are equal from Theorem 4.15. Since the hyperbolic sine is the same for supplementary angles, from Theorem 4.18 the triangles PTS_1 and PS_2T have proportional sides; therefore $t^2 = s_1 \cdot s_2$ follows. This product depends on the point and on hyperbola and it is called "The power of a point P with respect to hyperbola $\mathcal{H}$." The value of this product is given by (5.17)

and

$$s_1 \cdot s_2 = |(1 - m^2)(x_1 - x_p)(x_2 - x_p)| = |(1 - m^2)[x_1 x_2 - x_p(x_1 + x_2) + x_p^2]|$$

$$= \left| \frac{1 - m^2}{a}(a x_p^2 + b x_p + c) \right| \tag{5.14}$$

where, in the last passage, we use the link between the roots and the coefficients a, b, c of the equation of degree 2 (5.12), that has x_1, x_2 as roots. Substituting these coefficients, after reduction we obtain

$$s_1 \cdot s_2 = |x_p^2 - y_p^2 - 1|, \tag{5.15}$$

which is independent of m. Therefore the product (5.14) is the same for coincident solutions (tangent line) or if the intersection points are in the same or different arms of (5.10).   □

The "Euclidean demonstration" for the secant line crossing the different arms of an equilateral hyperbola is reported in the caption of Fig. 5.4.

The result of (5.15) can be generalized to hyperbolas with center in any point. We have

Theorem 5.9 *If from a non-external point* $P \equiv (x_p, y_p)$ *of an equilateral hyperbola*

$$(x - x_c)^2 - (y - y_c)^2 \pm p^2 = 0, \tag{5.16}$$

we trace a tangent line to the hyperbola, then the square of the distance between P *and the tangent point is obtained by substituting the coordinates of* P *in the equation for the hyperbola*

$$t^2 \equiv s_1 \cdot s_2 = |(x_c - x_p)^2 - (y_c - y_p)^2 \pm p^2|. \tag{5.17}$$

Proof The proof can be obtained, by means of the analytical method of Theorem 5.8 or by considering (5.16) instead of (5.10) and by noting that the last term in the round bracket of (5.14) can be obtained by substituting $x \rightarrow x_p$ in (5.12). This substitution, directly into (5.11), shows us that it must be $y = y_p$. Both these substitutions into (5.16) give (5.17).   □

We have [1, p. 195]

Theorem 5.10 *The value of the product* $\overline{PS_1} \cdot \overline{PS_2}$ *only depends on the point and on hyperbola. It is called* "The power of a point P with respect to hyperbola $\mathcal{H}$". *Its value can be obtained by means of* (5.17).

We note that in (5.16) we have collected, by taking $\pm p^2$, the four arms of the hyperbolas. These different signs give different values for the product $\overline{PS_1} \cdot \overline{PS_2}$, then it follows that the theorem holds only for secant lines crossing the arms of the same kind.

As a conclusion of this section we note that the theorems we have seen indicate that in some cases problems about equilateral hyperbolas may be solved more easily in the hyperbolic plane by applying the shown theory.

5.2 Triangle and Equilateral Hyperbolas

In this section we extend to equilateral hyperbolas the Euclidean theorems about circumcircle, incircle and ex-circles [2].

5.2.1 Circumscribed Hyperbola

As three points (in particular the vertexes of a triangle) in Euclidean plane define a circle, in the same way, in the hyperbolic plane, define an equilateral hyperbola (5.1), that has the same properties of Euclidean circle. We have

Theorem 5.11 *If we have three non-aligned points that can be considered the vertexes of a triangle, there is an equilateral hyperbola* (circumscribed hyperbola) *which passes through these points, and its semi-diameter is given by*

$$p = \frac{d_1 d_2 d_3}{4 S} \equiv \frac{d_n}{2 \sinh_e \theta_n}, \quad n = 1, 2, 3. \tag{5.18}$$

Proof With the usual analytical method we find its center $P_c \equiv (x_c, y_c)$ and the quadratic semi-diameter P and will see that the obtained result can be written as function of the lengths of the triangle sides.

Let us consider three non-aligned points $P_1 \equiv (x_1, y_1)$, $P_2 \equiv (x_2, y_2)$, $P_3 \equiv (x_3, y_3)$. The parametric equation of a hyperbola passing through them is obtained by imposing this condition on (5.1). By calling θ_i the hyperbolic angles corresponding to points P_i, we have

$$x_i = x_c \pm p \cosh_e \theta_i, \quad y_i = y_c \pm p \sinh_e \theta_i. \tag{5.19}$$

For finding the coordinates of the hyperbola center and the condition that states if it is of the first or second kind, we set the coordinate axes so that the points have the positions shown in Fig. 4.4 with coordinates stated in (4.54).

Since the center is on the axis of the chords given by the sides of the triangle, by considering the side $\overline{P_1 P_2}$, we have $x_c = x_2/2$. From the intersection between this one and another axis we obtain the center coordinates

$$P_c \equiv \left(\frac{x_2}{2}, \frac{y_3^2 - x_3^2 + x_2 x_3}{2 y_3} \right).$$

Moreover, since the hyperbola passes through the coordinate origin, the quadratic semi-diameter (P) is given by

$$P = x_c^2 - y_c^2 \equiv \frac{x_2^2 y_3^2 - (y_3^2 - x_3^2 + x_2 x_3)^2}{4 y_3^2} \equiv \frac{(x_3^2 - y_3^2)[y_3^2 - (x_2 - x_3)^2]}{4 y_3^2}. \tag{5.20}$$

Taking into account the last expressions of (4.34) and (4.56), we can write the denominator of (5.20) as $y_3^2 = (d_1 d_2 \sinh \theta_3)^2 / d_3^2 \equiv S^2 / D_3$ and, by means of (4.55), we see that also the numerator can be written as a function of invariant quantities

$$P = -\frac{D_1 D_2 D_3}{16 S^2}; \tag{5.21}$$

for $P > 0$ we have an equilateral hyperbola of the second kind, while for $P < 0$ we have an equilateral hyperbola of the first kind. Thus in relation to the hyperbola type we could say that there are two kinds of triangles depending on the sign of $D_1 \cdot D_2 \cdot D_3$.

By using Hero's formula (4.44), we obtain P as function of the sides quadratic lengths. If we set $p = \sqrt{|P|}$, from (5.21) and (4.34) we obtain (5.18), which is the same relation that holds for the radius of a circumcircle in a Euclidean triangle. □

For a triangle in a general position we can obtain, with a more laborious calculation [3, Chap. 6], the center coordinates:

$$z_c = \frac{[D_1 o(z_3 - z_2) + D_2 o(z_1 - z_3) + D_3 o(z_2 - z_1)]}{4 \cdot S}, \tag{5.22}$$

where D_{no}; $n = 1, 2, 3$ represents the quadratic distance of vertex "n" from the coordinate origin.

An example of circumscribed hyperbolas

Let us reconsider the triangle of Fig. 4.4, with a right angle in the vertex P_2 and determinate, as function of the sides lengths the kind of circumscribed hyperbolas.

From (5.21) and (4.55), we have $P = x_3^2 y_3^2 (x_3^2 - y_3^2)$, therefore

- if $|x_3| > |y_3| \Rightarrow P > 0 \Rightarrow$ second kind hyperbolas,
- if $|x_3| < |y_3| \Rightarrow P < 0 \Rightarrow$ first kind hyperbolas.

5.2.2 Inscribed and Ex-inscribed Hyperbolas

Let be given three points which can be considered the vertexes of a triangle, and the three straight lines given by the prolongation of the sides. We consider the problem of finding a hyperbola tangent to the triangle sides that, by analogy with the Euclidean problem about the circle, we call *inscribed hyperbola* and three hyperbolas tangent to the straight lines prolongations of the three sides that we call *ex-inscribed hyperbolas*. This problem represents an extension to hyperbolic geometry of a well-known Euclidean problem.

The solution of this problem shows the properties which are preserved when going from Euclidean to hyperbolic geometry and which properties must be considered peculiar to the Euclidean geometry. On the other hand we cannot construct a hyperbola inside a triangle, but we see that the analytical method allows us to solve together the problems concerning inscribed and ex-inscribed hyperbolas and to find the four hyperbolas.

Proof By utilizing a property of Euclidean incircles and excircles, we change the problem to the following one: *to find hyperbolas with their centers equidistant from three straight lines*.

This problem has a solution; moreover the hyperbolas which we find, also have other properties of the corresponding Euclidean circles.

We note that we must consider, in general, both conjugate arms of the hyperbolas, since the quadratic distance from a straight line and the center of a hyperbolas can be positive or negative depending on the sides (straight lines) kind, as we also see in a numerical example (Sect. 5.2.3).

For the solution of the problem we must find the center $P_c \equiv (x_c, y_c)$ and the quadratic semidiameter P of the hyperbolas. This problem can be solved by means of a linear system of two equations.

Actually from Theorem 3.3 we know that the distance between a point P_c and the straight line γ_E is a linear function of the coordinates of P_c, given by (3.18).

Let us consider the straight lines determined by the three points P_1, P_2, P_3 and let us denote γ_i the straight line between points $P_j, P_k, i \neq j, k$

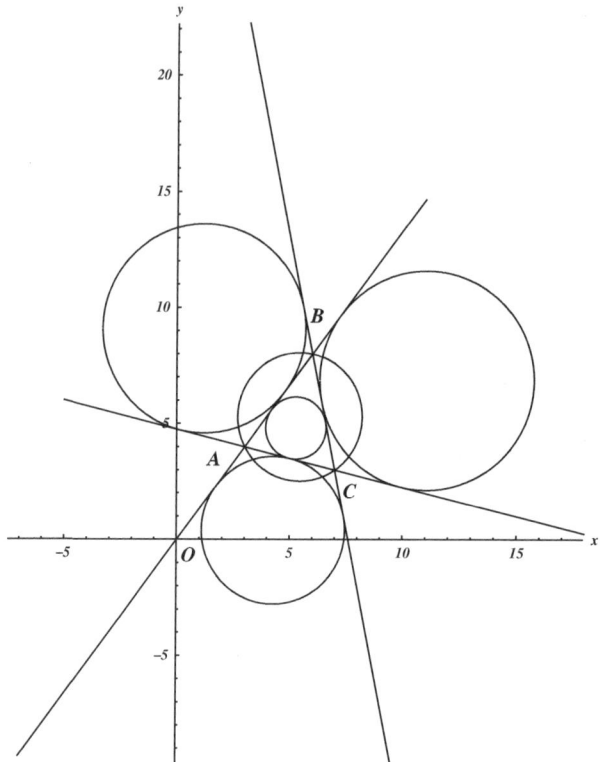

Fig. 5.5 Circumcircle, incircle and excircles of a Euclidean triangle. The incircle and excircles have the common property of having their centers equidistant from the sides or their continuation. In a Cartesian representation this condition can be formalized by means of two linear equations. Actually, taking into account that the distance between a point and a straight line have a plus or minus sign depending on the position of point with respect to straight line, we equate the distances but for the sign and, in this way we solve the problem by means of the same system of two linear equations. This approach allows us to calculate the radius and the center by means of (5.26) and (5.27) just by setting the Euclidean instead of hyperbolic distances

$$\gamma_i : \left\{ y - y_j = \frac{y_k - y_j}{x_k - x_j}(x - x_j) \mid i \neq j \neq k \right\}, \tag{5.23}$$

by calling $d_{c\gamma_i}$ the distance between $P_c \equiv (x_c, y_c)$ and γ_i, we have

$$d_{c\gamma_i} = \frac{(y_c - y_j)(x_k - x_j) - (y_k - y_j)(x_c - x_j)}{d_i}, \tag{5.24}$$

As in Euclidean geometry this quantity can be positive or negative, depending on the position of the point with respect to the straight line.

For the solution of the present problem we note that, as for Euclidean geometry, the centers of excircles and incircle, are on opposite sides with respect to one straight line. Therefore the four centers can be obtained by means of the same equations, by equating the distances by considering both the signs.

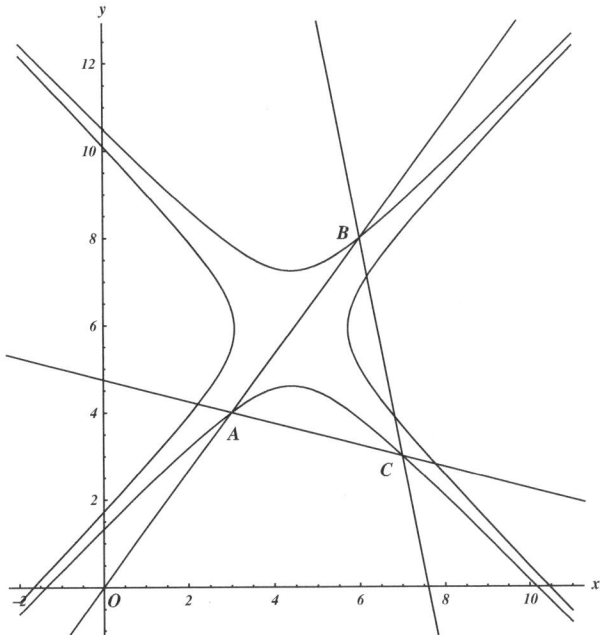

Fig. 5.6 Circumscribed hyperbola to a hyperbolic triangle. Let us consider the three points $A \equiv$ (3, 4), $B \equiv$ (6, 8), $C \equiv$ (7, 3) in the hyperbolic plane as the vertexes of a triangle. As in Euclidean plane we can associate with this triangle, and the straight lines given by the continuations of the sides, five hyperbolas. In this figure we represent the circumscribed hyperbola. Its center is obtained by (5.22), its quadratic semidiameter by (5.21). Since in this example D_{12}, $D_{13} < 0$, $D_{23} > 0$, we have the quadratic semidiameter $P < 0$, i.e., as one can note from the figure, the circumscribed hyperbola is of the first kind $(y - y_c)^2 - (x - x_c)^2 = k^2$. Also if the problem is solved by the arms of specific kind (5.21), in the figure we have reported all the four conjugate arms

By introducing two quantities ϵ_1, ϵ_2 equal to ± 1, we find the centers of the inscribed and circumscribed hyperbolas by solving the following four linear systems that differ just for the values of ϵ_1, ϵ_2:

$$d_{c\gamma_1} = \epsilon_1 \, d_{c\gamma_2}, \qquad d_{c\gamma_1} = \epsilon_2 \, d_{c\gamma_3}. \tag{5.25}$$

The semidiameters shall be obtained from the center coordinates and one of (5.24). $\square$

By solving the system (5.25), with $d_{c\gamma_i}$ given by (5.24) and setting $\epsilon_3 = \epsilon_1 \epsilon_2$, we obtain [2]

Theorem 5.12 *The centers and semidiameters of inscribed and ex-inscribed hyperbolas as functions of the side lengths of the triangle, are given by*

$$\begin{aligned}
x_c &= \frac{\epsilon_1 x_1 d_1 - \epsilon_2 x_2 d_2 + \epsilon_3 x_3 d_3}{\epsilon_1 d_1 - \epsilon_2 d_2 + \epsilon_3 d_3}, \\
y_c &= \frac{\epsilon_1 y_1 d_1 - \epsilon_2 y_2 d_2 + \epsilon_3 y_3 d_3}{\epsilon_1 d_1 - \epsilon_2 d_2 + \epsilon_3 d_3};
\end{aligned} \tag{5.26}$$

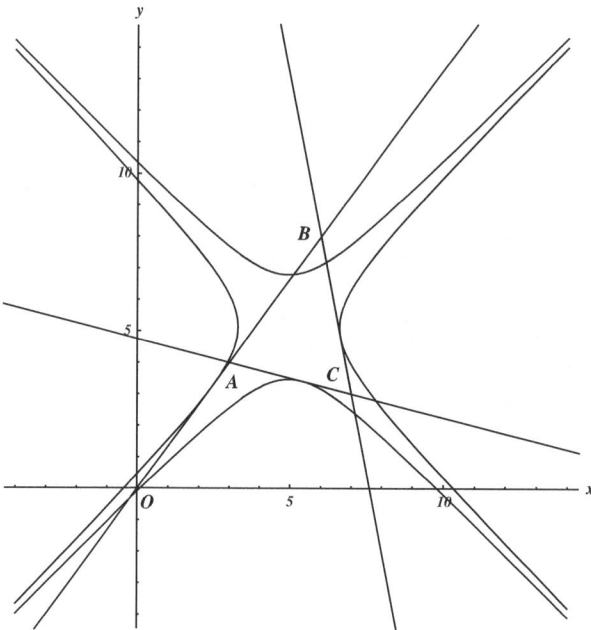

Fig. 5.7 Inscribed hyperbola. This problem and the next one for ex-inscribed hyperbolas is solved by means of the same linear system (5.25). Since the hyperbolas have the same kind of their tangent lines, the problem is, in general, solved by considering all their four arms. For our specific problem by the convention of Sect. 3.3; we have that the sides $\overline{AB}$ and $\overline{BC}$ (the straight lines γ_3, γ_1) are of the second kind, the side $\overline{AC}$ (the straight lines γ_2) is of the first kind. The hyperbolas tangent to γ_3 and γ_1 are of the second kind $(x - x_c)^2 - (y - y_c)^2 = k^2$, the hyperbola tangent to γ_2 is of the first kind $(x - x_c)^2 - (y - y_c)^2 = -k^2$. The center of hyperbola, inside the triangle, is given by (5.26), its semidiameter by (5.27). By analogy with Euclidean geometry, we define as inscribed the hyperbola with the smallest semidiameter by taking in (5.27) $\epsilon_1 = -\epsilon_2 = \epsilon_3 = -1$

$$p = \frac{2\,S}{|\epsilon_1\,d_1 - \epsilon_2\,d_2 + \epsilon_3\,d_3|} \tag{5.27}$$

where ϵ_1, $\epsilon_2 = \pm 1$ and we have also introduced $\epsilon_3 = \epsilon_1\,\epsilon_2$.

The centers and the semidiameters of inscribed and ex-inscribed hyperbolas are obtained by means of the same equations. By analogy with Euclidean geometry, we define as inscribed the *hyperbola* with the smallest semidiameter, i.e., $\epsilon_1 = -\epsilon_2 = \epsilon_3 = -1$.

We note that the expressions for p as a function of the side lengths are the same as those of the circles of Euclidean geometry. For our formalization on a Cartesian plane we have also reported in (5.26), the coordinates of the centers of hyperbolas.

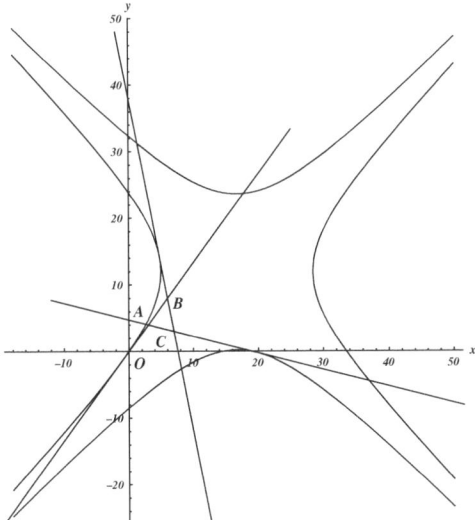

Fig. 5.8 Hyperbolas ex-inscribed to a triangle (case 1). In this figure and in the next ones, since the kinds of the hyperbolas are the same of the tangent lines they are the same as in Fig. 5.7. The centers of hyperbolas are given by (5.26), the semidiameter by (5.27), taking for ϵ_n the three values different from $\epsilon_1 = -\epsilon_2 = \epsilon_3 = -1$. As for ex-circle of Euclidean geometry, the ex-inscribed hyperbolas to an hyperbolic triangle have the property that their centers are external to triangle

5.2.3 Numerical Examples

In this example we show the five hyperbolas determined by a given triangle, and their graphical representation. Let us consider the triangle with vertexes in points $A \equiv (3, 4)$, $B \equiv (6, 8)$, $C \equiv (7, 3)$. In Euclidean plane a triangle, and the straight lines given by the continuations of the sides, define the circumcircle (passing through the vertexes), incircle (tangent to triangle sides) and three excircles (tangent to one side and to straight lines continuation of the other sides), represented in Fig. 5.5.

In hyperbolic plane the equilateral hyperbolas have the same properties of the circles. In Figs. 5.6, 5.7, 5.8 and 5.9 we represent the five hyperbolas.

For the circumscribed hyperbola the coordinates of the centers have been obtained by (5.22) and the quadratic semidiameters by (5.21). The centers and semidiameters of the inscribed and ex-inscribed hyperbolas are obtained by (5.26) and (5.27).

As far as the inscribed and ex-inscribed hyperbolas are concerned, their parameters are obtained by means of (5.26) and (5.27).

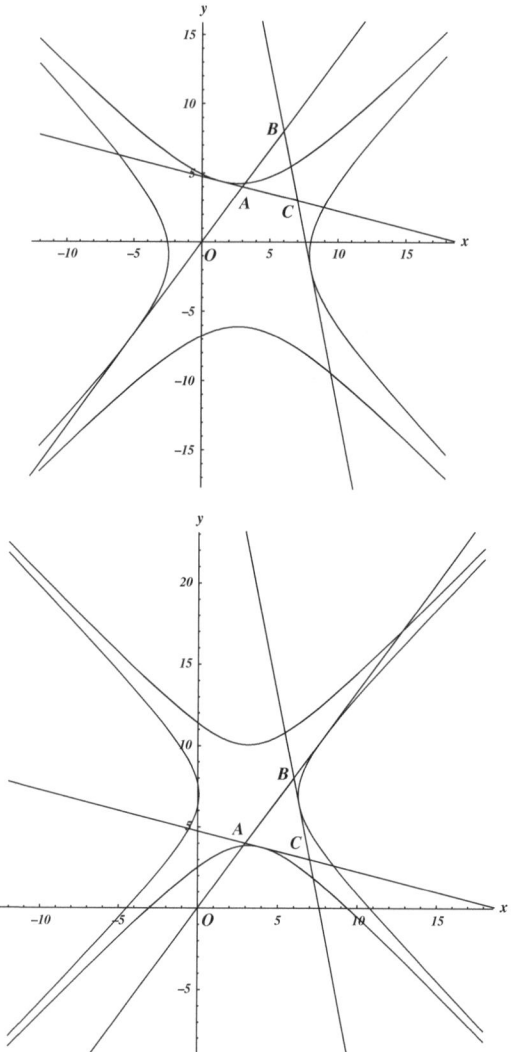

Fig. 5.9 Hyperbolas ex-inscribed to a triangle (cases 2 and 3)

References

1. I.M. Yaglom, *A Simple Non-euclidean Geometry and its Physical Basis*. (Springer-Verlag, New York, 1979)
2. F. Catoni, R. Cannata, V. Catoni, P. Zampetti, Two-dimensional hypercomplex numbers and related trigonometries and geometries. Adv. Appl. Clifford Al. **14**(1), 47 (2004)
3. F. Catoni, D. Boccaletti, R. Cannata, V. Catoni, E. Nichelatti, P. Zampetti, *The Mathematics of Minkowski Space-Time*. (Birkhäuser Verlag, Basel, 2008)

Chapter 6
The Motions in Minkowski Space–Time
(*Twin Paradox*)

Abstract All the curves in space-time plane, can be considered as a motion and the lengths of the time-like lines give the proper time. Therefore we can compare in a geometrical way, on different curves, the differences between proper times. Proper times are obtained by means of elementary mathematics for uniform, uniformly accelerated motions and their compositions; by means of differential calculus for the arbitrary time-like curves.

Keywords Proper time in space-time · Twin paradox for uniform motions · Twin paradox for uniformly accelerated motions · Twin paradox for general motions

In this chapter we show how the formalization of trigonometry in the pseudo-Euclidean plane allows us to treat exhaustively all kinds of motions and to give a complete formalization to what is today called "twin paradox." After a century this problem continues to be the subject of many papers, not only relative to experimental tests [1, 2] but also regarding physical and epistemological considerations [3]. We begin by recalling how this "name" originates.

The final part of Sect. 4 of Einstein's famous 1905 special relativity paper [4] contains sentences concerning moving clocks on which volumes have been written: ◁...If we assume that the result proved for a polygonal line is also valid for a continuously curved line, we obtain the theorem: If one of two synchronous clocks at A is moved in a closed curve with constant velocity until it returns to A, the journey lasting t seconds, then the clock that moved runs $\frac{1}{2} t \left(v/c\right)^2$ seconds[1] slower than the one that remained at rest. ▷

About 6 years later, on 10 April 1911, at the Philosophy Congress at Bologna, Paul Langevin replaced the clocks A and B with human observers and the "twin paradox" officially was born. Langevin, using the example of a space traveler who travels a distance L (measured by someone at rest on the Earth) in a straight line to

---

[1] Obviously neglecting magnitudes of fourth and higher order.

a star in one year and then abruptly turns around and returns on the same line, wrote:

◁... Revenu à la Terre ayant vielli deux ans, il sortira de son arche et trouvera notre globe vielli deux cents ans si sa vitesse est restée dans l'interval inférieure d'un vingt-millième seulement à la vitesse de la lumière.[2] ▷

Apart from the "humanization" that complicates the acceptance of this consequence of Special Relativity, another contradiction results from the first Einstein's postulate. Actually, from this postulate, if two reference systems move, one relatively to the other, in uniform motion, there are not physical experiments that allow us to state which one is in motion. As a consequence each of the twins regard himself as stationary and the other one in motion. Following a Galileo's example: a passenger on a boat along a coast considers the earth as moving. Therefore each of the twins must see the other as younger.

We must remark that Langevin himself stresses the point which will be the subject of subsequent discussions, that is the asymmetry between the two reference frames. The space traveler undergoes acceleration through his journey for starting, halfway and for stopping while the twin at rest in the Earth reference frame always remains in an inertial frame.

These accelerations, without a specific quantification, are usually considered as the reason for the time differences. Recent experiments [1, 2] have belied this hypothesis.

We think that a conclusion on the role of accelerated motions and, most of all, an evaluation of the amount of slowing down of accelerated frames, can be reached through the mathematics of special relativity. The self-consistency of the formalization of hyperbolic trigonometry in previous chapters, allows us to solve problems for every motion in the Minkowski space–time as if we were working on the Euclidean plane [5, 6].

We note that in some examples, even if we could obtain the result by using directly the hyperbolic equivalent of Euclidean theorems, we have preferred not to use this approach. Actually, the results are obtained by means of elementary mathematics and the obtained correspondences represent alternative demonstrations with respect to the ones given in previous chapters.

6.1 Inertial Motions

In a representative (t, x) plane let us start with the following example:

- The first twin is steady at the point $x = 0$, his path is represented by the t axis.

---

[2] Langevin's address to the Congress of Bologna was published in *Scientia* **10**, 31–34 (1911). As reported by Miller [4], the popularization of relativity theory for philosophers had an immediate impact which we can know from the comment of one of the philosophers present. Henry Bergson (1922) wrote: ◁...it was Langevin's address to the Congress of Bologna on 10 April 1911 that first drew our attention to Einstein's ideas. We are aware of what all those interested in the theory of relativity owe to the works and teachings of Langevin." ▷

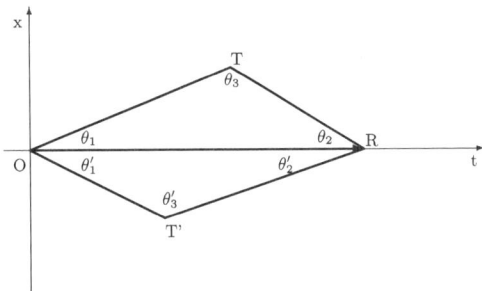

Fig. 6.1 The twin paradox for uniform (inertial) motions. The twin paradox is usually formalized in the particular example of a twin stationary and the other one moving with constant speed and an inversion of the speed direction up to meeting again. In this figure we represent arbitrary uniform motions for both the twins. The results of Chaps. 4 and 5 allow us the complete formalization of this problem. We begin by considering the motion represented by the triangle OTR with $\theta_1 \neq \theta_2$. (1) The first twin is stationary in $x = 0$, its motion is then represented by t axis. (2) The second twin starts with constant velocity v_1 from point $O \equiv (0, 0)$ and in point T, he change the velocity direction and came back with constant velocity v_2, and arrives in point R. From the geometrical point of view we can compare the elapsed times for the twins by means of the hyperbolic lengths (proper times) of the side $\overline{OR}$ with the sum of sides $\overline{OT} + \overline{TR}$. By means of the relation between velocities and hyperbolic angles (2.32) we have $\tanh \theta_1 = v_1$ and $\tanh \theta_2 = v_2$, respectively and we set $\widehat{OTR} = \theta_3$. By calling p the semidiameter of circumscribed equilateral hyperbola to triangle OTR, by means of (5.18), we have $\overline{OT} = 2p \sinh \theta_2$; $\overline{TR} = 2p \sinh \theta_1$; $\overline{OR} = 2p \sinh \theta_3$, then $\frac{\tau_I}{\tau_{II}} = \frac{\overline{OR}}{\overline{OT} + \overline{TR}} = \frac{\sinh \theta_3}{\sinh \theta_2 + \sinh \theta_1} = \frac{\sinh[\theta_1 + \theta_2]}{\sinh \theta_2 + \sinh \theta_1}$. The last relation, in which θ_3 is given as function of the known hyperbolic angles (or velocity), is obtained by applying (4.41), (5.18) and the formula (4.23). Actually, we have $\overline{OR} = \overline{OT} \cosh \theta_1 + \overline{TR} \cosh \theta_2 \Rightarrow \sinh \theta_3 = \cosh \theta_1 \sinh \theta_2 + \cosh \theta_2 \sinh \theta_1 \equiv \sinh[\theta_1 + \theta_2]$. For both twins in motion, by comparing $\overline{OT} + \overline{TR}$ and $\overline{OT'} + \overline{T'R}$ with $\overline{OR}$, we obtain $\frac{\tau}{\tau'} = \frac{\sinh[\theta_1 + \theta_2]}{\sinh[\theta_1' + \theta_2']} \frac{\sinh \theta_2' + \sinh \theta_1'}{\sinh \theta_2 + \sinh \theta_1}$

- The second twin, on a rocket, starts with constant speed v_1 from $O \equiv (0, 0)$ and after a time τ_1, at the point T, he reverses its direction and comes back with constant speed v_2, arriving at the point $R \equiv (\tau_2, 0)$. From a physical point of view the speed cannot change in a null time, but this time can be considered negligible with respect to τ_1, τ_2. An experimental result with only uniform motion is reported in [1, 2]. In this experiment the lifetime of the muon in the CERN muon storage ring was measured.

In Fig. 6.1. we represent this problem by means of the triangle OTR.

Solution From a geometrical point of view we can compare the elapsed travel times for the twins by comparing the "lengths" (proper times) of the sum $\overline{OT} + \overline{TR}$ with the side $\overline{OR}$.

Now we see that the Euclidean formalization of space–time trigonometry allows us to obtain a simple quantitative formulation of the problem.

By means of (2.32) we call $\theta_1 \equiv \tanh^{-1} v_1$ the hyperbolic angle $\widehat{ROT}, \theta_2 \equiv \tanh^{-1} v_2$ the hyperbolic angle $\widehat{ORT}$ and θ_3 the hyperbolic angle $\widehat{OTR}$. Given their physical meaning, the angles θ_1 and θ_2 are such that the straight lines $\overline{OT}$ and $\overline{TR}$ are time-like [7] (in a Euclidean representation the angle of straight lines with the t axis must be less than $\pi/4$). Applying the law of cosines (4.41) to $\overline{OR}$; we have

$$\overline{OR} = \overline{OT} \cosh \theta_1 + \overline{TR} \cosh \theta_2. \tag{6.1}$$

By calling τ_I and τ_{II} the proper times of the twins and $\tau_I - \tau_{II} = \Delta\tau$, it follows that the difference between the twins' proper times $\Delta\tau$ is

$$\Delta\tau \equiv \overline{OR} - \overline{OT} - \overline{TR} = \overline{OT}(\cosh \theta_1 - 1) + \overline{TR}(\cosh \theta_2 - 1) > 0. \tag{6.2}$$

If we call p the semi-diameter of the equilateral hyperbola circumscribed to triangle $\widehat{OTR}$, from (5.18), we have

$$\overline{OT} = 2p \sinh \theta_2; \quad \overline{TR} = 2p \sinh \theta_1; \quad \overline{OR} = 2p \sinh \theta_3, \tag{6.3}$$

and (6.2), becomes

$$\begin{aligned}\Delta\tau &= 2p(\cosh \theta_1 \sinh \theta_2 + \cosh \theta_2 \sinh \theta_1 - \sinh \theta_1 - \sinh \theta_2) \\ &\equiv 2p[\sinh(\theta_1 + \theta_2) - \sinh \theta_1 - \sinh \theta_2]. \end{aligned} \tag{6.4}$$

In a similar way we have

$$\frac{\tau_I}{\tau_{II}} = \frac{\overline{OR}}{\overline{OT} + \overline{TR}} = \frac{\sinh \theta_3}{\sinh \theta_2 + \sinh \theta_1} = \frac{\sinh[\theta_1 + \theta_2]}{\sinh \theta_2 + \sinh \theta_1}. \tag{6.5}$$

The last relation can be obtained by means of the relations among the angles of triangle (4.58) or by (4.41) and (6.3) as it is shown in the caption of Fig. 6.1.□

Resetting in (6.5) the hyperbolic trigonometric functions as function of the speeds by means of (2.34), after reduction, we have

$$\frac{\tau_I}{\tau_{II}} = \frac{v_1 + v_2}{v_1 \sqrt{1 - v_2^2} + v_2 \sqrt{1 - v_1^2}}. \tag{6.6}$$

Equations 6.4 and 6.5 allow us to calculate $\Delta\tau$ or τ_I/τ_{II} for every specific problem.

Now we consider the following problem: given $\theta_1 + \theta_2 = \text{const} \equiv 2\,\theta_m$, what is the relation between θ_1 and θ_2, so that $\Delta\tau$ has its greatest value?

Proof The straightforward solution is

$$\begin{aligned} \Delta\tau &= 2p[\sinh 2\,\theta_m - \sinh \theta_1 - \sinh(2\,\theta_m - \theta_1)], \\ \frac{d(\Delta\tau)}{d\,\theta_1} &= 0 \Rightarrow \theta_1 = \theta_m \equiv \theta_2, \\ \frac{d^2(\Delta\tau)}{d\,\theta_1^2} \bigg|_{\theta_1 = \theta_m} &= -\sinh \theta_m < 0. \end{aligned} \tag{6.7}$$

□

We have obtained the "intuitive Euclidean" solution that the greatest difference between elapsed times, i.e., the shortest proper-time for the moving twin, is obtained for $\theta_1 = \theta_2$. For this value (6.5) corresponds to the well-known solution [4]

$$\tau_{\overline{OR}} = \tau_{\overline{(OT+TR)}} \cosh \theta_1 \equiv \frac{\tau_{\overline{(OT+TR)}}}{\sqrt{1-v^2}}. \tag{6.8}$$

Now we give a geometrical interpretation of this problem. From (4.58) we know that if $\theta_1 + \theta_2 = 2\theta_m$, with $\theta_m = $ const, then θ_3 is constant too, then the posed problem is equivalent to: what can be the position of vertex T if starting and final points and angle θ_3 are given?

The problem is equivalent to having, in a triangle, a side and the opposite angle. In an equivalent problem in Euclidean geometry we know that the vertex T does move on a circle's arc. Then, from the established correspondence (Theorem 5.6) of circles in Euclidean geometry with equilateral hyperbolas in pseudo-Euclidean geometry, we have that in the present space–time problem the vertex T moves on an arc of an equilateral hyperbola.

Now let us generalize the twin paradox to the case in which both twins change their state of motion: their motions start in O, both twins move on different straight lines and cross again in R. The graphical representation is given by a quadrilateral figure and we call the other two vertexes T and T'. Since a hyperbolic rotation does not change hyperbolic angles between the sides and hyperbolic side lengths (Theorem 4.14), we can rotate the figure so that the vertex R lies on the t axis (see Fig. 6.1). The problem can be considered as a duplicate of the previous one in the sense that we can compare proper times of both twins with side $\overline{OR}$. If we indicate by $(')$ the quantities referred to the triangle under the t axis, we apply (6.5) twice and obtain τ and τ' for every specific example

$$\frac{\tau}{\tau'} = \frac{\sinh[\theta_1 + \theta_2]}{\sinh \theta_2 + \sinh \theta_1} \frac{\sinh \theta_2' + \sinh \theta_1'}{\sinh[\theta_1' + \theta_2']}. \tag{6.9}$$

In particular, if we have $\theta_1 + \theta_2 = \theta_1' + \theta_2' = 2\theta_m$, from the result of (6.7) it follows that the youngest twin is the one for which θ_1 and θ_2 are closer to θ_m.

6.2 Inertial and Uniformly Accelerated Motions

Now we consider some "more realistic" examples in which uniformly accelerated motions are taken into account. Actually in the representative plane t, x all the regular curves with time-like tangent lines can be considered representing the motion of a body. In particular let us consider a curve written, as function of the parameter θ, as

$$t = t(\theta), \quad x = x(\theta). \tag{6.10}$$

The parameter θ can be arbitrarily taken otherwise, as we better see in the following of this section and in Sect. 6.3, it is more convenient to take as natural parameter the proper time on the curve which can be obtained from (6.10):

$$\tau = \int_0^P \sqrt{\left(\frac{dt}{d\theta}\right)^2 - \left(\frac{dx}{d\theta}\right)^2} \, d\theta \Rightarrow \frac{d\tau}{d\theta} = \sqrt{\left(\frac{dt}{d\theta}\right)^2 - \left(\frac{dx}{d\theta}\right)^2}. \tag{6.11}$$

Setting (6.10) as function of τ the relativistic velocity and the relativistic acceleration are introduced as the vectors with components

$$v_t \equiv \dot{t} = \frac{dt}{d\tau}, \quad v_x \equiv \dot{x} = \frac{dx}{d\tau}; \tag{6.12}$$

$$a_t \equiv \dot{v}_t = \frac{d^2 t}{d\tau^2}, \quad a_x \equiv \dot{v}_x = \frac{d^2 x}{d\tau^2}. \tag{6.13}$$

We know that in classical kinematics the velocity and acceleration are three-dimensional vectors. In special relativity, for the stated equivalence between time and space, the four-dimensional velocity and acceleration are introduced. Now we note that the corresponding spatial components of three and four-dimensional velocity are different whereas the corresponding components of three and four-dimensional acceleration are the same [7, Sect. 1.4]. The same considerations hold for the only spatial component we are here considering.

Now we show

Theorem 6.1 *In a natural parametrization the relativistic velocity is a unit vector.*

Proof By considering (6.10) as function of τ, we have

$$v_t \equiv \frac{dt}{d\tau} = \frac{dt}{d\theta} \frac{d\theta}{d\tau}, \quad v_x \equiv \frac{dx}{d\tau} = \frac{dx}{d\theta} \frac{d\theta}{d\tau}, \tag{6.14}$$

therefore with a natural parametrization, by means of (6.11) we have:

$$|\mathbf{v}| \equiv \left(\sqrt{\left(\frac{dt}{d\theta}\right)^2 - \left(\frac{dx}{d\theta}\right)^2}\right) \frac{d\theta}{d\tau} = 1, \tag{6.15}$$

$\square$

These general motions will be considered in Sect. 6.3, here we apply the exposed considerations to the characteristic curves of hyperbolic plane, i.e., the equilateral hyperbolas given, in parametric form, by (5.1)

$$\mathcal{I} \equiv \begin{cases} t = t_C \pm p \sinh \theta \\ x = x_C \pm p \cosh \theta \end{cases} \quad \text{for } -\infty < \theta < +\infty. \tag{6.16}$$

For these hyperbolas $C \equiv (t_C, x_C)$ represents the center and θ is a parameter that, from a geometrical point of view, represents a hyperbolic angle measured with respect to an axis passing through C and parallel to the x axis (Sect. 4.2). In (6.16) the $+$ sign refers to the upper arm and the $-$ sign to the lower one.

We also have

$$dt = \pm p \cosh \theta \, d\theta, \quad dx = \pm p \sinh \theta \, d\theta, \tag{6.17}$$

and the proper time on the hyperbola

$$\tau_{\mathcal{I}} = \int_0^\theta \sqrt{dt^2 - dx^2} \equiv \int_0^\theta p \, d\theta \equiv p \, \theta. \tag{6.18}$$

This relation states the link between proper time, semidiameter and hyperbolic angle and also shows that hyperbolic angles are given by the ratio between the "lengths" of hyperbola arcs and semi-diameter, as circular angles in Euclidean trigonometry are given by the ratio between circle arcs and radius. Moreover, as shown in Sect. 4.5.1 the magnitude of hyperbolic angles can be calculated in a Euclidean way.

Now let us see the kinematic properties of the "hyperbolic motion." By taking the proper time as parameter, the (6.16) become

$$\mathcal{I} \equiv \begin{cases} t = t_C \pm p \, \sinh[p^{-1}\tau] \\ x = x_C \pm p \, \cosh[p^{-1}\tau]. \end{cases} \tag{6.19}$$

From (6.19) we obtain the components of relativistic velocity and acceleration

$$v_t \equiv \frac{dt}{d\tau} = \cosh[p^{-1}\tau], \quad v_x \equiv \frac{dx}{d\tau} = \sinh[p^{-1}\tau]; \tag{6.20}$$

$$a_t \equiv \frac{d^2 t}{d\tau^2} = p^{-1} \sinh[p^{-1}\tau], \quad a_x \equiv \frac{d^2 x}{d\tau^2} = p^{-1} \cosh[p^{-1}\tau]. \tag{6.21}$$

From (6.20) we can check (6.15): $|v| = 1$.

From (6.21) we have

$$|a| \equiv \sqrt{|a_t^2 - a_x^2|} = p^{-1}. \tag{6.22}$$

The equilateral hyperbolas in the t, x plane represent uniformly accelerated motions (in the classical meaning of the term, since a is the same as the "classical" acceleration). Therefore all equilateral hyperbolas with a given diameter correspond to a motion with constant acceleration $a = p^{-1}$. These motions are the ones we consider in this Section together with uniform motions.

Coming back to (6.16), at the point P, determined by $\theta = \theta_1$, we have $v \equiv dx/dt = \tanh \theta_1$ and the straight line, tangent to the hyperbola for $\theta = \theta_1$, is given by (Chap. 4):

$$x - (x_C \pm p \cosh \theta_1) = \tanh \theta_1 [t - (t_C \pm p \sinh \theta_1)]$$
$$\Rightarrow x \cosh \theta_1 - t \sinh \theta_1 = x_C \cosh \theta_1 - t_C \sinh \theta_1 \mp p. \tag{6.23}$$

From this equation we see that θ_1 also represents the hyperbolic angle of the tangent to hyperbola with t axis. Taking into account that the angle θ_1 that defines the point on the hyperbola is measured with respect to a straight line parallel to x axis while θ_1 in (6.23) is measured with respect to t axis, the straight lines determined by these angles are symmetric with respect to axes bisectors, i.e., the semi-diameter $\overline{CP}$ is pseudo-orthogonal to the tangent in P (see also Figs. 3.2 and 6.3). This property corresponds, in Euclidean geometry, to the well-known property of a circle where the radius is orthogonal to the tangent-lines.

6.2.1 Uniform and Uniformly Accelerated Motions

We start with the following example in which the first twin, after some accelerated and decelerated motions with the same acceleration (p^{-1}), returns to the starting point with vanishing velocity. The problem is represented in Fig. 6.2.

- The first twin (I) starts with a constant acceleration (represented with the hyperbola $\mathcal{I}_1$) from O to A and then a constant decelerated motion up to V with $v(V) = 0$ and then accelerated with reversed velocity up to $A', (v(A') = v(A))(\mathcal{I}_2)$, then another decelerated motion ($\mathcal{I}_3$) up to $B \equiv (4\,t_A, 0)$, $(v(B) = 0)$.
- The second twin (II) moves with a uniform motion ($\mathcal{T}_1$) that, without loss of generality, can be represented as stationary at the point $x = 0$.

Solution Hyperbola $\mathcal{I}_1$ has its center at $C \equiv (0, -p)$. Then its equation is given by

$$\mathcal{I}_1 \equiv \begin{cases} t = p \sinh \theta \\ x = p(\cosh \theta - 1) \end{cases} \quad \text{for } 0 < \theta < \theta_1, \qquad (6.24)$$

and $A \equiv (p \sinh \theta_1,\ p \cosh \theta_1 - p)$.

The symmetry of the problem indicates that for both twins the total elapsed times are four times the elapsed times of the first motion. The proper time of twin I is obtained from (6.18),

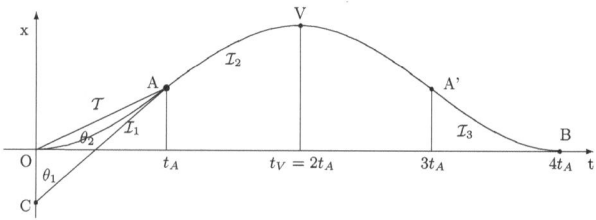

Fig. 6.2 Uniform and uniformly accelerated motions. The proper time for a uniformly accelerated motion is compared with the proper time of two uniform motion: the first one stationary (represented by the t axis); the second one represented by the line $\overline{OA}$ (Sect. 6.2.1)

$$\tau_I \equiv 4\,\tau_{I_1} = 4 \int\limits_0^{\theta_1} \sqrt{d\,t^2 - d\,x^2} \equiv 4 \int\limits_0^{\theta_1} p\,d\theta \equiv 4p\theta_1, \qquad (6.25)$$

the proper time of twin II is four times the abscissa of point A,

$$\tau_{II} \equiv 4\,t_A = 4p\,\sinh\theta_1. \qquad (6.26)$$

The difference between the proper times is $\Delta\tau = 4p(\sinh\theta_1 - \theta_1)$, and their ratio is

$$\frac{\tau_I}{\tau_{II}} = \frac{\theta_1}{\sinh\theta_1}. \qquad (6.27)$$

Now we consider the motion $\mathcal{T}$ on the side $\overline{OA}$.

Solution Let us call θ_2 the hyperbolic angle between straight line OA and the t axis; the equation of the straight line OA is

$$\mathcal{T} \equiv \{x = t\,\tanh\theta_2\} \qquad (6.28)$$

and we calculate θ_2, imposing that this straight line crosses the hyperbola (6.24) for $\theta = \theta_1$. By substituting (6.28) in (6.24), and by using from the second and the third passage (4.28) and (4.26), we have

$$\begin{cases} t = p\,\sinh\theta_1 \\ t\,\tanh\theta_2 = p(\cosh\theta_1 - 1) \end{cases} \Rightarrow \frac{\sinh\theta_2}{\cosh\theta_2} = \frac{\cosh\theta_1 - 1}{\sinh\theta_1} \Rightarrow \theta_1 = 2\,\theta_2, \quad (6.29)$$

This result can be considered a new demonstration of Theorem 5.4, i.e., the central angle (θ_1) is twice the hyperbola angle (θ_2) on the same chord. Then we have

$$\overline{OA} = \frac{\overline{Ot_A}}{\cosh\theta_2} \equiv \frac{p\,\sinh\theta_1}{\cosh\theta_2} \equiv 2p\,\sinh\theta_2 \qquad (6.30)$$

and, taking into account the proper time on the hyperbola given by (6.25), we obtain

$$\frac{\tau_{\mathcal{I}}}{\tau_{\mathcal{T}}} = \frac{\theta_2}{\sinh\theta_2}. \qquad (6.31)$$

$\square$

Relation 6.31 has a simple "Euclidean" interpretation. Actually it can be interpreted by means of the correspondence (Chap. 5) with Euclidean geometry, where it represents the ratio between the length of a circle arc and its chord. We have to note how, in Euclidean geometry, this ratio is greater than 1 whereas in hyperbolic geometry it is less than 1, as the reverse triangle inequality requires. We shall look into this property for curved lines in Sect. 6.2.3.

We can verify that the ratio between (6.27) and (6.31), is the ratio (6.8).

6.2.2 Uniform Motions Joined by Uniformly Accelerated Motions

In this example we connect the two sides of the triangle of Fig. 6.1. by means of an equilateral hyperbola, i.e., we consider the decelerated and accelerated (with reversed velocity) motions too (Fig. 6.3).

- The twin I moves from $O \equiv (0, 0)$ to $P \equiv (p \sinh \theta_1, p \cosh \theta_1 - p)$ with a uniform motion, indicated as $\mathcal{T}_1$, then he goes on with a constant decelerated motion up to V, and then he accelerates with reversed velocity up to P', where he has the same velocity as the initial one, and moves again with uniform velocity up to $R \equiv (t_R, 0)(\mathcal{T}_1')$.
- The twin II moves with a uniform motion ($\mathcal{T}_2$) which, without loss of generality, can be represented as stationary in the point $x = 0$.

Solution A mathematical formalization can be the following: let us consider the decelerated and accelerated motions represented by the equilateral hyperbola (6.16) for $-\theta_1 < \theta < \theta_1$ and the tangent to the hyperbola for $\theta = \theta_1$ as given by (6.23). The straight line (6.23) in parametric form is given by

$$\mathcal{T}_1 \equiv \begin{cases} t = \tau \cosh \theta_1 \\ x = \tau \sinh \theta_1 \end{cases}, \qquad (6.32)$$

where τ is the proper time on the straight line.

For the considerations after (6.23), the angle PCV is given by θ_1, and the proper time from P to the vertex V is $\tau_{\mathcal{I}} = p\theta_1$.

Thanks to the symmetry of the problem, the total proper times from O to R is a duplicate of these ones.

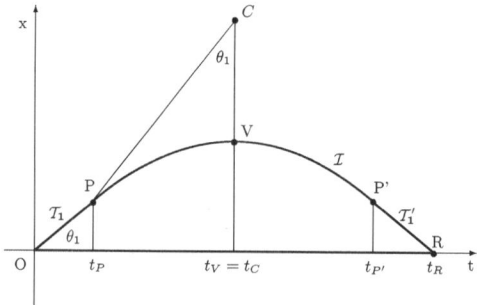

Fig. 6.3 Uniform motions joined by uniformly accelerated motions. The sides OT and TR of the triangle in Fig. 6.1. are taken equal and are joined by an hyperbola representing a motion decelerated and accelerated up to obtaining the starting speed with inverted direction. We will see that if the time of decelerated and accelerated motions are less with respect to uniform motions, we obtain (6.2) or the equivalent (6.8)

For twin II we have: $\tau_{II} \equiv 2\,t_C = 2(\tau \cosh\theta_1 + p \sinh\theta_1)$, therefore we have

$$\Delta\tau = 2[\tau(\cosh\theta_1 - 1) + p(\sinh\theta_1 - \theta_1)]. \qquad (6.33)$$

$\square$

The proper time on this rounded off triangle is greater than the one on the triangle, as we shall better see in the next example. The physical interpretation is that the velocity on the hyperbola arc is less than the one on straight lines OT, TR of Fig. 6.1.

We can check at once that if the time of decelerated and accelerated motions are small with respect to uniform motions, the (6.33) is the same as (6.2) or the equivalent (6.8), in which we set $\overline{OT} = \overline{TR}$, i.e., $\theta_1 = \theta_2$.

6.2.3 Reversed Triangle Inequality for Curved Lines

In the following example we consider the motions

1. stationary motion in $x = 0$;
2. on the upper triangle of Fig. 6.1. with sides $\overline{OT} = \overline{TR}$;
3. on hyperbola $\mathcal{I}$ tangent in O and in R to sides $\overline{OT}$ and $\overline{TR}$, respectively;
4. on hyperbola $\mathcal{I}_c$ circumscribed to triangle OTR.

The problem is represented in Fig. 6.4. In this example we expose a formalization of the triangle inequality, reversed with respect to the Euclidean plane (Sect. 4.6.1) applied to both straight lines and curved lines, in particular to equilateral hyperbolas. We shall see that the shorter the lines (trajectories) look in a Euclidean representation (Fig. 6.4), the longer they are in space–time geometry (6.35).

Solution Side $\overline{OT}$ lies on the straight line represented by the equation

$$x \cosh\theta_1 - t \sinh\theta_1 = 0. \qquad (6.34)$$

Hyperbola $\mathcal{I}$ is obtained requiring that it be tangent to straight line (6.34) in O. We obtain from (6.16) and (6.23) $t_C = p \sinh\theta_1$, $x_C = p \cosh\theta_1$ and, from the definitions of hyperbolic trigonometry, $\overline{OT} = t_C/\cosh\theta_1 \equiv p \tanh\theta_1$.

For hyperbola $\mathcal{I}_c$, circumscribed to the triangle OTR, its semi-diameter is given by (5.18): $p_c = \overline{OT}/(2 \sinh\theta_1) \equiv p/(2 \cosh\theta_1)$. If we call C_c its center and $2\,\theta_c$ the angle $\widehat{OC_cR}$, we note that θ_c is a central angle of chord $\overline{OT}$ while θ_1 is a hyperbola angle on the chord $\overline{TR} = \overline{OT}$. Then, as has been shown in Sect. 6.2.1, we have $\theta_c = 2\,\theta_1$.

Summarizing the lengths (proper times) for the motions are:

1. $\overline{OR} \equiv 2\,t_T = 2p \sinh\theta_1$;
2. from (6.34) it follows that

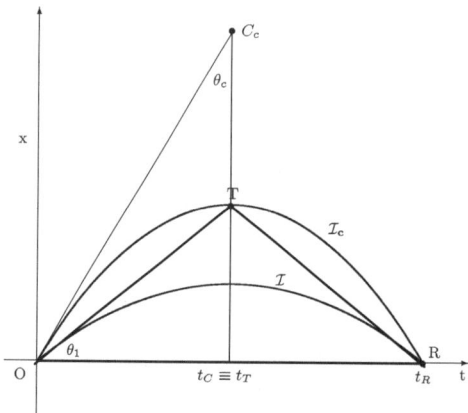

Fig. 6.4 Twin paradox and reversed triangle inequality for curved lines. In this example we expose a generalization of the triangle inequality, reversed with respect to the Euclidean plane, (Sect. 4.6.1) applied to both straight lines and equilateral hyperbolas. In particular we consider the uniform motions: • stationary in $x = 0$: represented by the segment $\overline{OR}$; • on the triangle OTR, with sides $\overline{OT} = \overline{TR}$ (velocity $v_1 = \tanh \theta_1$); and uniformly accelerated motions from O to vertex of hyperbolas and decelerated from vertex to R; • on the hyperbola $\mathcal{I}$ tangent in O and R to sides $\overline{OT}$ and $\overline{TR}$, respectively; • on the hyperbola $\mathcal{I}_c$ circumscribed to triangle OTR. By writing $\mathrm{arc}(\mathcal{I})$ and $\mathrm{arc}(\mathcal{I}_c)$ the hyperbolic lengths of hyperbola arcs, we obtain: $\overline{OR} \equiv 2p \sinh \theta_1 > \mathrm{arc}(\mathcal{I}) \equiv 2p\theta_1 > \overline{OT} + \overline{TR} \equiv 2p \tanh \theta_1 > \mathrm{arc}(\mathcal{I}_c) \equiv 2p\theta_1 / \cosh \theta_1$. Therefore, as shorter the lines (trajectories) look in a Euclidean representation as longer they are in space–time geometry

$$T \equiv (p \sinh \theta_1, \, p \sinh \theta_1 \tanh \theta_1), \text{ so that } \overline{OT} = \overline{TR} = p \tanh \theta_1;$$

3. from (6.18) the length of arc of hyperbola $\mathcal{I}$ between O and R is given by

$$\tau_{\mathcal{I}} = 2p\theta_1;$$

4. from (6.18) the length of arc of $\mathcal{I}_c$ from O and R is given by

$$\tau_{\mathcal{I}_c} = 2p_c\theta_c \equiv 2p\theta_1 / \cosh \theta_1.$$

Then the following relations hold:

$$\begin{aligned}
\overline{OR} \equiv 2p \sinh \theta_1 > \; & \mathrm{arc}(\mathcal{I}) \equiv 2p\theta_1 > \overline{OT} + \overline{TR} \\
\equiv 2p \tanh \theta_1 > \; & \mathrm{arc}(\mathcal{I}_c) \equiv 2p\theta_1 / \cosh \theta_1.
\end{aligned} \tag{6.35}$$

$\square$

As a corollary of this example we consider the following problem: given the side $\overline{OR} = \tau$ (proper time of the stationary twin) what is the proper time of twin I moving on an equilateral hyperbola, as a function of the acceleration p^{-1}?

The answer is: *as acceleration p^{-1} does increase, proper time $\tau_{\mathcal{I}}$ can be as little as we want.*

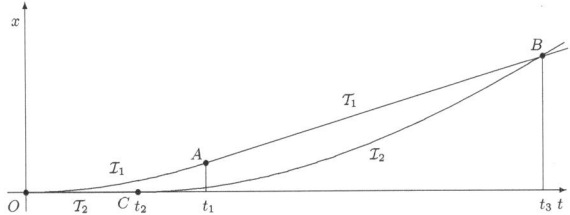

Fig. 6.5 The most general uniform and accelerated motions. The most general motion, in which both the twins after uniform and accelerated motions, meet again. This general result (6.41) can be applied to specific cases

Proof From the hyperbolic motion of (6.16) we have $t = \tau/2 - p \sinh \theta$ and for $t = 0$ we have $\theta = \theta_1$, therefore $2p \sinh \theta_1 = \tau$, and for relativistic motions $(\theta_1 \gg 1)$, we obtain

$$\tau \simeq p \exp[\theta_1] \Rightarrow \theta_1 \simeq \ln\frac{\tau}{p}. \tag{6.36}$$

Then from relation (6.18), $\tau_{\mathcal{I}} \equiv 2p\theta_1 = 2p \ln[\tau/p]_{\longrightarrow}^{p\to 0} 0$. $\qquad\qquad\square$

6.2.4 The Most General Uniform and Accelerated Motions

We conclude this section with a more general example in which both twins have a uniform and an accelerated (with the same acceleration) motion.

- The first twin (I) starts with a constant accelerated motion and then goes on with a uniform motion.
- The second twin (II) starts with a uniform motion and then goes on with a constant accelerated motion.

The problem is represented in Fig. 6.5.

Solution We can represent this problem in the (t, x) plane in the following way:

(I) starts from point $O \equiv (0, 0)$ with an acceleration given by $p^{-1}(\mathcal{I}_1)$ up to point A (time t_1), then goes on with a uniform motion $(\mathcal{T}_1)$ up to point B in which meet the twin (II) (time t_3). The mathematical expression of $\mathcal{I}_1$ is given by (6.24), and $A \equiv (p \sinh \theta_1;\ p \cosh \theta_1 - p)$,

(II) starts from the point $O \equiv (0, 0)$ with a uniform motion $(\mathcal{T}_2)$, (stationary in $x = 0$) up to point C in a time $t_2 = \alpha p \sinh \theta_1$ which we have written proportional to t_1. Then he goes on with an accelerated motion $(\mathcal{I}_2)$, with the same acceleration p^{-1} of twin (I) up to crossing the trajectory of twin (I) at time t_3.

The analytical representation of $\mathcal{I}_1$ is given by (6.24). $\mathcal{I}_2$ is represented by

$$\mathcal{I}_2 \equiv \begin{cases} t = p(\alpha \sinh \theta_1 + \sinh \theta) \\ x = p(\cosh \theta - 1) \end{cases} \text{ for } 0 < \theta < \theta_2, \qquad (6.37)$$

where θ_2 represents the value of the hyperbolic angle in the crossing point B between $\mathcal{I}_2$ and $\mathcal{T}_1$.

$\mathcal{T}_1$ is given by the straight line tangent to $\mathcal{I}_1$ in θ_1:

$$\mathcal{T}_1 \equiv \{x \cosh \theta_1 - t \sinh \theta_1 = p(1 - \cosh \theta_1)\}. \qquad (6.38)$$

From (6.37) and (6.38) we calculate the crossing point between $\mathcal{I}_2$ and $\mathcal{T}_1$. We have

$$\cosh(\theta_2 - \theta_1) = \alpha \sinh^2 \theta_1 + 1. \qquad (6.39)$$

This equation has an explicit solution for $\alpha = 2$. Actually we have

$$\cosh(\theta_2 - \theta_1) = 2 \sinh^2 \theta_1 + 1 \equiv \cosh 2\theta_1 \Rightarrow \theta_2 = 3\theta_1.$$

Let us calculate the proper times.

- The proper times relative to the accelerated motions are obtained from (6.18) $\tau_{\mathcal{I}_1} = p\theta_1$, $\tau_{\mathcal{I}_2} = p\theta_2$.
- The proper time relative to $\mathcal{T}_2$ is given by $t_2 = \alpha p \sinh \theta_1$.
- On straight line $\mathcal{T}_1$, between A and $B \equiv p(\alpha \sinh \theta_1 + \sinh \theta_2, \cosh \theta_2 - 1)$, the proper time is obtained by means of hyperbolic trigonometry,

$$\tau_{\mathcal{T}_1} \equiv \overline{AB} = (x_B - x_A)/\sinh \theta_1 \equiv p(\cosh \theta_2 - \cosh \theta_1)/\sinh \theta_1. \qquad (6.40)$$

Then the complete proper times of the twins are

$$\begin{aligned} \tau_I &= p[\theta_1 + (\cosh \theta_2 - \cosh \theta_1)/\sinh \theta_1] \\ \tau_{II} &= p(\alpha \sinh \theta_1 + \theta_2). \end{aligned} \qquad (6.41)$$

$\square$

Let us consider relativistic velocities ($v = \tanh \theta_i \simeq 1 \Rightarrow \theta_1, \theta_2 \gg 1$); in this case we can approximate the hyperbolic functions in (6.39) and (6.41) with the positive exponential term and, for $\alpha \neq 0$, we obtain from (6.39): $\exp[\theta_2 - \theta_1] \simeq \alpha \exp[2\theta_1]/2$, and from (6.41)

$$\tau_I \simeq p(\theta_1 + \exp[\theta_2 - \theta_1]) \simeq p(\theta_1 + \alpha \exp[2\theta_1]/2), \quad \tau_{II} \simeq p(\alpha \exp[\theta_1]/2 + \theta_2). \qquad (6.42)$$

The greatest contributions to the proper times are given by the exponential terms that derive from uniform motions. If we neglect the linear terms with respect to the exponential ones, we obtain a ratio of the proper times *independent of the $\alpha \neq 0$ value*,

$$\tau_I \simeq \tau_{II} \exp[\theta_1]. \qquad (6.43)$$

The twin that moves for a shorter time with uniform motion has the shortest proper time. Since the total time is the same, a shorter time with uniform motion means a longer time with accelerated motion, so this result is the same as (6.31).

With regard to the result of this example we could ask: how is it possible that a uniform motion close to a light-line is the longer one?

We can answer to this question by having a glance at (6.40). Actually in this equation the denominator $\sinh \theta_1 \gg 1$ takes into account that the motion is close to a light-line, but in the numerator $\cosh \theta_2 \gg \cosh \theta_1$ indicates that the crossing point B is so far away that its contribution is the determining term of the result we have obtained.

6.3 Non-uniformly Accelerated Motions

This section has been inserted for giving an exhaustive formulation of all the problems. In any case we have to specify that, as a difference with the previous chapters where the problem have been solved by means of elementary geometry, some notion of advanced mathematics are now necessary. For this reason, before to formalizing the general motions in the hyperbolic plane we recall some definitions and relations about the curves in Euclidean plane.

6.3.1 The Curves in Euclidean Plane

Let us consider in a Cartesian representation a regular curve given by the parametric equations

$$x = x(u), \quad y = y(u). \tag{6.44}$$

To every value of the parameter u in an interval corresponds a point P. As positive direction on the curve we take the one in which the point moves as u increases.

On this curve the velocity and the acceleration are defined by means of the following relations

$$v_x \equiv \dot{x} = \frac{dx}{du}, \quad v_y \equiv \dot{y} = \frac{dy}{du}; \tag{6.45}$$

$$a_t \equiv \dot{v}_t = \frac{d^2 x}{du^2}, \quad a_y \equiv \dot{v}_y = \frac{d^2 y}{du^2}. \tag{6.46}$$

These names derive from the classical kinematics. Actually if for a curve represented in a Euclidean plane or space we consider the parameter u as the time, (6.44) represent the motion of a point for which the definitions (6.45) and (6.46) are the velocity and the acceleration, respectively.

The parameter u can be taken in an arbitrary way otherwise, for the reasons we are going to see, it can be convenient to take the **natural parameter** representing the arc length on the curve from a fixed point (the origin) up to the considered point P. In particular we have

$$s = \int_0^P \sqrt{\left(\frac{dx}{du}\right)^2 + \left(\frac{dy}{du}\right)^2} \, du \ \Rightarrow \ \frac{ds}{du} = \sqrt{\left(\frac{dx}{du}\right)^2 + \left(\frac{dy}{du}\right)^2}. \tag{6.47}$$

If the parametric equations (6.44) are written as functions of s, we have

$$v_x \equiv \frac{dx}{ds} = \frac{dx}{du}\frac{du}{ds}, \quad v_y \equiv \frac{dy}{ds} = \frac{dy}{du}\frac{du}{ds}, \tag{6.48}$$

then

$$|\mathbf{v}| \equiv \left(\sqrt{\left(\frac{dx}{du}\right)^2 + \left(\frac{dy}{du}\right)^2}\right) \frac{du}{ds} = 1. \tag{6.49}$$

From a geometrical point of view, with this parametrization, v_x e v_y represent the direction cosines of tangent to the curve. We also have

Theorem 6.2 *In a natural parametrization the velocity and acceleration vectors are orthogonal.*

Proof From (6.49), by differentiating the first and last terms, we have

$$\frac{d|\mathbf{v}|^2}{ds} \equiv \frac{d|\mathbf{v}\cdot\mathbf{v}|}{ds} \equiv 2\,(\mathbf{v}\cdot\dot{\mathbf{v}}) \equiv 2\,(\mathbf{v}\cdot\mathbf{a}) = 0. \tag{6.50}$$

$\square$

Therefore by calling $\mathbf{n}$ the unitary vector of the normal to the curve, we can write

$$\frac{d\mathbf{v}}{ds} = K(s)\,\mathbf{n}. \tag{6.51}$$

Let us now see the meaning of $K(s)$: from a geometrical point of view, the derivative of $\mathbf{v}$ gives the rate of variation of the direction of the tangent, i.e., how it draws away from a straight direction. The modulus of this quantity is called **curvature** and is indicated with $K(s)$. Therefore $K(s) = \sqrt{\ddot{x}^2 + \ddot{y}^2}$.

Let us now see its value in the simplest cases:

- For a straight line, from (6.44), we have $x = as$, $y = bs$, therefore $K(s) = 0$.
- For a circle, from (6.44), we have $x = R\cos[R^{-1}s]$, $y = R\sin[R^{-1}s]$, from which $K(s) = R^{-1}$, is constant.

For a general curve $K(s)^{-1}$ represents the radius of the circle that better approximate the curve in the considered point. This circle is called osculating circle.

We note that the curvature depends on the second derivate with respect to s, therefore it is independent of the direction of s fixed on the curve.

From the stated definitions and the properties found, an important theorem follows

Theorem 6.3 *For the curves represented by means of the natural parameter s, the following* Frenet formulas *hold*

$$\frac{d\mathbf{v}}{ds} = K(s)\,\mathbf{n}, \tag{6.52}$$

$$\frac{d\mathbf{n}}{ds} = -K(s)\,\mathbf{v}. \tag{6.53}$$

Proof Equation 6.52 corresponds to (6.51).

The demonstration of (6.53) is straightforward: actually since $\mathbf{n}$ is a unitary vector (6.50) holds true. Therefore its derivative has the direction of $\mathbf{v}$ and, as for (6.51) we can write

$$\frac{d\mathbf{n}}{ds} = \alpha\,\mathbf{v}. \tag{6.54}$$

Taking into account that $\mathbf{v} \cdot \mathbf{n} = 0$, from (6.52) and (6.54) we obtain α and (6.53):

$$\frac{d}{ds}(\mathbf{v} \cdot \mathbf{n}) \equiv \left(\frac{d\mathbf{v}}{ds} \cdot \mathbf{n}\right) + \left(\frac{d\mathbf{n}}{ds} \cdot \mathbf{v}\right) \equiv K + \alpha = 0 \Rightarrow \alpha = -K. \tag{6.55}$$

$\square$

Now we see

Theorem 6.4 *The Frenet formulas allow us to obtain the parametric equations of a curve if we know the curvature $K(s)$.*

Proof For this formalization we use complex numbers and this will allow us to extend these results to pseudo-euclidean plane by using hyperbolic numbers.

Calling $\phi(s)$ the angle between the tangent-line to the curve and the x axis, we can write:

$$v_x \equiv \frac{dx}{ds} = |\mathbf{v}|\cos\phi(s) \equiv \cos\phi(s); \quad v_y \equiv \frac{dy}{ds} = |\mathbf{v}|\sin\phi(s) \equiv \sin\phi(s). \tag{6.56}$$

Then $\mathbf{v}$ can be written in the formalism of complex numbers:

$$\mathbf{v} \equiv v_x + i\,v_y = \exp[i\,\phi(s)] \tag{6.57}$$

and the orthogonal unit vector **n** (Sect. 3.3), as

$$\mathbf{n} = \mathrm{i}\,\exp[\mathrm{i}\,\phi(s)]. \tag{6.58}$$

Substituting (6.57) and (6.58) into (6.52), we obtain

$$\frac{d\,\exp[\mathrm{i}\,\phi(s)]}{d\,s} \equiv \{\mathrm{i}\,\exp[\mathrm{i}\,\phi(s)]\}\frac{d\,\phi(s)}{d\,s} = K(s)\{\mathrm{i}\,\exp[\mathrm{i}\,\phi(s)]\}$$

$$\Rightarrow \quad \phi(s) = \phi_0 + \int K(s)\,d\,s. \tag{6.59}$$

The same result could be obtained by means of (6.53).

By means of the value of $\phi(s)$ from (6.59), we obtain the components of **v** from (6.57). Writing the curve (6.44), in the language of complex vectors, we have

$$z(s) = x(s) + \mathrm{i}\,y(s), \tag{6.60}$$

and, by integrating (6.48), we obtain the coordinates

$$x + \mathrm{i}\,y \equiv \int [v_x + \mathrm{i}\,v_y]d\,s = \int \exp[\mathrm{i}\,\phi(s)]\,d\,s. \tag{6.61}$$

By the value of $K(s)$ from 6.59, we obtain the result we are looking for

$$x(s_A) = x_0 + \int_0^{s_A} \cos\phi(s')\,d\,s' \equiv t_0 + \int_0^{s} \cos\left[\int_0^{s'} K(s'')\,d\,s''\right]d\,s', \tag{6.62}$$

$$y(s_A) = y_0 + \int_0^{s_A} \sin\phi(s')\,d\,s' \equiv x_0 + \int_0^{s} \sin\left[\int_0^{s'} K(s'')\,d\,s''\right]d\,s'. \tag{6.63}$$

$$\square$$

6.3.2 Formalization of Non-uniformly Accelerated Motions

In this section we see that the formalism introduced in previous chapters and the correspondence between complex and hyperbolic numbers allows us to extend the results shown for the Euclidean plane to the lines that can represent a motion in the t, x plane (i.e., with time-like tangent lines).

This motion is, in general, non-uniform and can be considered as the envelope of the equilateral hyperbolas with semidiameter corresponding to the instantaneous accelerations. Or, vice versa, we can construct in every point of a curve an "osculating hyperbola" which has the same properties of the osculating circle in Euclidean geometry. Actually the semi-diameters of these hyperbolas are linked to the second derivative with respect to the line element as the radius of osculating circles in Euclidean geometry.

In the representative plane t, x let us represent points P by means of the hyperbolic coordinates and consider a time-like curve from the origin of the coordinates to point A which, without loss of generality (see Sect. 6.2) we write by means of the proper time as parameter

$$t = t(\tau), \quad x = x(\tau) \tag{6.64}$$

$$\mathbf{v} = \mathbf{v}(\tau); \quad v_t \equiv \dot{t} = \frac{dt}{d\tau}, \quad v_x \equiv \dot{x} = \frac{dx}{d\tau}. \tag{6.65}$$

Now we consider the problem: to calculate the proper time in an inertial frame if we know the acceleration on the curve as a function of the proper time on the curve. The problem is equivalent to the one solved by means of complex numbers in Euclidean geometry (6.62) and (6.63). Therefore, thanks to the established correspondence between Euclidean and space–time geometry, the problem can be solved in two steps by means of hyperbolic numbers.

As a first step we state Frenet's formulas for the space–time plane.

Frenet's formulas in the Minkowski space–time.

Let us consider a parametrization by means of proper time: we have

Theorem 6.5 *In the space–time plane the following* Frenet-like formulas *hold*:

$$\frac{d\mathbf{v}}{d\tau} = K(\tau)\,\mathbf{n}, \tag{6.66}$$

$$\frac{d\mathbf{n}}{d\tau} = K(\tau)\,\mathbf{v}, \tag{6.67}$$

where $\mathbf{n}$ *represents the unit vector pseudo-orthogonal to the curve,* $K(\tau)$ *the modulus of acceleration and, from* (6.22), $1/K(\tau)$ *the semi-diameter of osculating hyperbolas.*

We have changed the symbol p, introduced in Chap. 4, with $K(\tau)$, that recalls the non-constant curvature of lines in the Euclidean plane.

Proof If the parameter τ is the proper time, $\ddot{t}$ and $\ddot{x}$ are the components of a vector $\mathbf{a}$ that represents the relativistic acceleration. Since from (6.15) we have that $\mathbf{v}$ is a unit vector, by indicating with a dot $(\cdot)$ the scalar product in space–time (Sect. 4.2), as for Euclidean plane (6.50), we have

$$\frac{d\,|\mathbf{v}|^2}{d\tau} \equiv \frac{d\,|\mathbf{v}\cdot\mathbf{v}|}{d\tau} \equiv 2(\mathbf{v}\cdot\dot{\mathbf{v}}) \equiv 2(\mathbf{v}\cdot\mathbf{a}) = 0 \tag{6.68}$$

Then, $\mathbf{v}$ is orthogonal to $\mathbf{a}$, that is a space-like vector. By calling $\mathbf{n}$ the unit vector orthogonal to the curve, we can write the proportionality equation (6.66). Equation (6.66) is equivalent to the first Frenet formula in the Euclidean plane (6.52). Now we show that also the second Frenet formula holds. Actually, we have

$$\mathbf{n} \cdot \mathbf{n} = -1 \Rightarrow \mathbf{n} \cdot \frac{d\mathbf{n}}{d\tau} = 0; \qquad (6.69)$$

as for Eq. 6.66 we can write

$$\frac{d\mathbf{n}}{d\tau} = \alpha \mathbf{v} \qquad (6.70)$$

and taking into account that $\mathbf{v} \cdot \mathbf{n} = 0$, from (6.66) and (6.70) we obtain α,

$$\frac{d}{d\tau}(v \cdot n) \equiv \left(\frac{dv}{d\tau} \cdot \mathbf{n}\right) + \left(\frac{d\mathbf{n}}{d\tau} \cdot \mathbf{v}\right) = 0 \equiv -K + \alpha \Rightarrow \alpha = K. \qquad (6.71)$$

So we have the second Frenet-like formula (6.67).   □

Now we see

Theorem 6.6 *Frenet space–time formulas allow us to find the proper time in an inertial frame by knowing the "curvature" (acceleration) of the curve as function of the proper time on the curve.*

Proof Calling $\theta(\tau)$ the hyperbolic angle between the tangent-line to the curve and t axis, we can write

$$v_t \equiv \frac{dt}{d\tau} = |\mathbf{v}| \cosh \theta(\tau) \equiv \cosh \theta(\tau); \quad v_x \equiv \frac{dx}{d\tau} = |\mathbf{v}| \sinh \theta(\tau) \equiv \sinh \theta(\tau).$$
$$(6.72)$$

Then $\mathbf{v}$, written in "hyperbolic polar" form, is given by

$$\mathbf{v} \equiv v_t + \mathrm{h}\, v_x = \exp[\mathrm{h}\, \theta(\tau)] \qquad (6.73)$$

and the pseudo-orthogonal unit vector $\mathbf{n}$ (Sect. 3.3), by

$$\mathbf{n} = \mathrm{h}\, \exp[\mathrm{h}\, \theta(\tau)]. \qquad (6.74)$$

Substituting (6.73) and (6.74) into (6.66), we obtain

$$\frac{d \exp[\mathrm{h}\, \theta(\tau)]}{d\tau} \equiv \{\mathrm{h}\, \exp[\mathrm{h}\, \theta(\tau)]\} \frac{d\theta(\tau)}{d\tau} = K(\tau)\{\mathrm{h}\, \exp[\mathrm{h}\, \theta(\tau)]\}$$
$$\Rightarrow \theta(\tau) = \theta_0 + \int K(\tau)\, d\tau. \qquad (6.75)$$

The same result could be obtained by means of (6.67).

By means of the value of $\theta(\tau)$ obtained by (6.75), we obtain the components of $\mathbf{v}$ and $\mathbf{n}$, from (6.73) and (6.74), respectively.

By integrating (6.65), we have

$$t + \mathrm{h}\, x \equiv \int [v_t + \mathrm{h}\, v_x]\, d\tau = \int \exp[\mathrm{h}\, \theta(\tau)]\, d\tau. \qquad (6.76)$$

By means of the value of $\theta(\tau)$ from (6.75) we obtain, from (6.76), the expressions equivalent to (6.62) and (6.63).

$$t(\tau_A) = t_0 + \int_0^{\tau_A} \cosh \theta(\tau') \, d\tau' \equiv t_0 + \int_0^{\tau_A} \cosh \left[\int_0^{\tau'} K(\tau'') \, d\tau'' \right] d\tau', \qquad (6.77)$$

$$x(\tau_A) = x_0 + \int_0^{\tau_A} \sinh \theta(\tau') \, d\tau' \equiv x_0 + \int_0^{\tau_A} \sinh \left[\int_0^{\tau'} K(\tau'') \, d\tau'' \right] d\tau'. \qquad (6.78)$$

$\square$

In Sect. 6.2 we have seen that in the representative (t, x) plane the equilateral hyperbola represents a uniformly accelerated motion. Now we can complete this result by means of

Theorem 6.7 *The only motions with constant acceleration are the hyperbolic motions.*

Proof The parametric equations of a motion with constant acceleration are obtained from (6.77) and (6.78) by setting $K(\tau) = $ const. $\equiv p^{-1}$. Therefore by solving two elementary integrals we obtain (6.19). $\square$

6.3.3 Proper Time in Non-uniformly Accelerated Motions

Now we return to the initial problem by comparing the proper times in the following motions (see Fig. 6.6)

1. motion I, a non-uniformly accelerated motion of which we know the acceleration as a function of the proper time on the curve $(K(\tau))$,
2. motion II, stationary in $x = 0$,
3. motion III, uniform motion from O to A. The straight line OA forms with the t axis the hyperbolic angle $\theta = \tanh^{-1}[x(\tau_A)/t(\tau_A)]$.

We have

1. (I) proper time, τ_A,
2. (II) proper time, $t(\tau_A)$ from (6.77),
3. (III) proper time $\tau(\overline{OA}) = t(\tau_A)/\cosh\theta$.

Motions (I) and (III) have the same starting and final points, therefore we can compare directly their proper times.

The motions (I) + (II) are two independent motions with different final points. Now we see, by the following considerations, that the final points can be set coincident. Actually since the curvature (acceleration) is given by the second derivative of t and x with respect to τ, it is independent of motion direction. Then

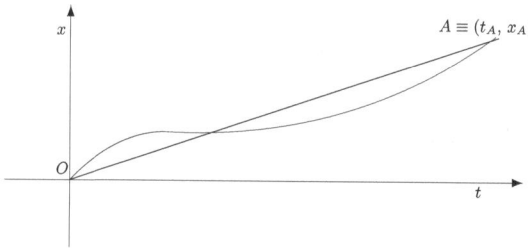

Fig. 6.6 Non-uniformly accelerated motions. In this figure we represent a problem that, in a future we do not know how far, is the answer to a "practical problem": space travelers who measure the time τ on their spaceship, how can they know the elapsed time $t(\tau)$ on the Earth? The solution of this problem is obtained, by means of hyperbolic numbers, extending to space–time the classical theory of curves in Euclidean plane (Sect. 6.3.2). Actually this time can be calculated by means of (6.77), knowing the speed (or the acceleration) of the rocket as function of the time (proper time) on the rocket: $\frac{\tau_{(I)}}{\tau_{(II)}} = \frac{\tau_A}{t(\tau_A)}$ where $t(\tau_A) = \int_0^{\tau_A} \cosh\left[\int_0^{\tau'} K(\tau'')\, d\tau''\right] d\tau'$

let us consider the curve obtained by reflecting (I) with respect to a straight line passing through A and parallel to x axis, and call O' the point in which this line crosses the t axis. The proper time on this curve is τ_A. Since also the time from O to O' for twin (II) is doubled, the total result is to double both the times of the "real" trip, obtaining the ratio

$$\frac{\tau_{(I)}}{\tau_{(II)}} = \frac{\tau_A}{t(\tau_A)}, \tag{6.79}$$

where $t(\tau_A)$ is given by (6.77).

Equation 6.79 allows to a space traveler, to construct an inertial clock (e.g., the time on the Earth) from data taken on its spaceship.

6.4   Conclusions

As we know, the twin paradox spreads far and wide having been considered the most striking exemplification of the space–time "strangeness" of Einstein's theory of special relativity. What we have striven to show is that hyperbolic trigonometry supplies us with an easy tool by which one can deal with any kinematic problem in the context of special relativity. Otherwise it allows us to obtain the *quantitative solution* of any problem and dispels all doubts regarding the role of acceleration in the flow of time.

Actually if we consider "true", for a lot of experimental confirmations, the space–time symmetry stated by Lorentz transformations, in whatever way obtained, the hyperbolic numbers provide the right mathematical structure inside which the two-dimensional problems must be dealt with. Finally, we remark that the application of hyperbolic trigonometry to relativistic space–time turns out to be a "Euclidean way" of dealing with the pseudo-Euclidean plane.

References

1. J. Bailey et al., Precise measurement of the anomalous magnetic moment of the muon, Nuovo Cimento A **9**, 369 (1972)
2. J. Bailey et al., Precise measurement of the anomalous magnetic moment of the muon, Nature **268**, 301 (1977)
3. E. Minguzzi, Differential aging from acceleration, an explicit formula, Am. J. Phys. **73**(9), 876 (2005) and the references therein
4. A. Einstein, We are quoting from Miller's translation of the German text of *On the electrodynamics of moving bodies*, English translation published as an appendix of: Arthur I. Miller, *Albert Einstein's special theory of relativity: Emergence (1905) and Early Interpretation (1905–1911)*, Springer (1998)
5. D. Boccaletti, F. Catoni, V. Catoni, The "Twins Paradox" for uniform and accelerated motions. Adv. Appl. Clifford Al. **17**(1), 1 (2007); The "Twins Paradox" for non-uniformly accelerated motions. Adv. Appl. Clifford Al. **17**(4), 611 (2007)
6. F. Catoni, D. Boccaletti, R. Cannata, V. Catoni, E. Nichelatti, P. Zampetti, *The Mathematics of Minkowski Space-Time* (Birkhäuser Verlag, Basel, 2008)
7. G.L. Naber, *The Geometry of Minkowski Spacetime. An Introduction to the Mathematics of the Special Theory of Relativity, Sect. 1.4* (Springer, New York, 1992)

Chapter 7
Some Final Considerations

The modern view of geometry, that we recall in the Appendix, has allowed us to formalize the "geometry of space–time" so that we can work in this plane as we usually do for Euclidean plane geometry. Otherwise the obtained mathematical system, following Euclidean geometry, combine the logical vision with the intuitive vision allowing us to agree with the following Einstein's thought [1].

◁ We honour the ancient Greece as the cradle of Western science. In this place a logical system, wonderment of thinking, has been firstly created. Their terms so clearly follow one from the others and no one of the demonstrations produce some doubts: we are speaking about Euclidean geometry.

This admirable work of human mind has given to men the greatest confidence for future efforts. ▷

Perhaps, thanks to this admiration of Euclid's Work, Einstein has formalized special relativity, starting from two axioms and applying the axiomatic-deductive method [2] of Euclidean geometry. Actually the transformation equations of special relativity seem to be a detachment from Euclidean geometry as it happened in the previous century with the non-Euclidean geometries. Otherwise the effort for stating a correspondence with Euclidean geometry by introducing an imaginary time is both misleading and mathematically wrong. Actually, as we have recalled in this book, the group that can be related with complex numbers is not the Lorentz's one but rather it is the same of Euclidean geometry (Sect. 2.1).

However as well as in nineteenth century the Euclidean and non-Euclidean geometries have been set into the same frame, so happens for Euclidean and space–time geometries, as we have shown in this book (Chap. 2) by relating these geometries with two systems of numbers that derive from the kind of solutions of a second degree equation (Sect. 2.2), Eq. (2.20). In particular:

1. From square roots of negative numbers we have complex numbers and Euclidean geometry.
2. From square roots of positive numbers we have hyperbolic numbers and Minkowski space–time geometry.

F. Catoni et al., *Geometry of Minkowski Space–Time*, SpringerBriefs in Physics, DOI: 10.1007/978-3-642-17977-8_7, © Francesco Catoni 2011

These geometries have a common origin and therefore can be considered as equivalent.

Perhaps the "genius intuitions" about the Nature's laws allowed Einstein to see an astonishing equivalence in spite of the apparent differences.

In any case if we want to have some hope of success of our "future efforts", we have to consider the "philosophical thinkings" of Einstein as relevant as his scientific results.

About these "philosophical thinkings" we recall two of them.

1. ◁ If we observe some coincidences in the Nature we must look for their deeper meaning. ▷
2. ◁ Our experience teaches us that the nature represents the realization of what we can imagine of most mathematically simple. I believe that a purely mathematical construction allows us to reveal the concepts which can give us the key for understanding the natural phenomena and the principles that link them together.

   Obviously the experimental confirmation is the only way for verifying a mathematical construction describing physical phenomena; but just in the mathematics we can find the creative principle [1]. ▷

Thanks to these principles we have obtained the "little results" shown in this book:

1. To look for the reason that has allowed complex numbers to become relevant for applied sciences notwithstanding they were introduced as "imaginary numbers".
2. To find how the generalization of the "mathematics of complex numbers" *creates* the formalization of the natural link between space and time.

As a final conclusion we point out another relevant consequence of the obtained formalization: once the relativistic effects have been properly explained as a new geometry, the understanding of gravity as a manifestation of curvature comes naturally and much easier. In particular a relation between curvature and accelerated motions in two-dimensional space–time has been obtained as a "geometrical" result, without the Einstein's equivalence principle [3].

Moreover, in recent papers [4, 5] the integration of General Relativity equations and also the determination of the constants of motion has been obtained by just geometrical methods.

In these papers the usual approach to General Relativity has been reversed: actually for obtaining the motions of test particle along geodesic lines, i.e., for integrating the General Relativity equations, the Newton conservation laws are generally used. In these papers the equations of geodesics are obtained by means of a method deriving from differential geometry and, from their equations, the constants of motion (conservation laws) follow.

References

1. A. Einstein, *On the Method of Theoretical Physics* (Clarendon Press, Oxford, 1934)
2. A. Einstein, We are quoting from Miller's translation of the German text of On the electrodynamics of moving bodies, English translation published as an appendix of: Arthur I. Miller, Albert Einstein's special theory of relativity: Emergence (1905) and Early Interpretation (1905–1911), (Springer, 1998)
3. F. Catoni, R. Cannata, V Catoni, P. Zampetti, Lorentz surface with constant curvature and their physical interpretation. Nuovo Cimento. B **120 B**(1), 37 (2005). Reprinted in F. Catoni, D. Boccaletti, R. Cannata, V. Catoni, E. Nichelatti, P. Zampetti, *The Mathematics of Minkowski Space-Time.* (Birkhäuser Verlag, Basel, 2008) Chaps.9, 10
4. D. Boccaletti, F. Catoni, R. Cannata, P. Zampetti, Integrating the geodesic equations in the Schwarzschild and Kerr space-times using Beltrami's "geometrical" method. Gen. Relat. Gravit. **37**(12), (2005)
5. F. Catoni, R. Cannata, P. Zampetti, "Geometrical" determination of the constants of motion in general relativity. Nuovo Cimento. B **124 B**(9), 975 (2009)

Appendix A: The Present-day Meaning of *Geometry*

The 19th century is today known as the century of industrial revolution but equally important is the revolution in mathematics and physics, that took place in 18th and 19th centuries. Actually we can say that the industrial revolution arises from the important technical and scientific progresses.

As far as the mathematics is concerned a lot of new arguments appeared, and split a matter that, for more than twenty centuries, has been represented by Euclidean geometry also considered, following Plato and Galileo, as the language for the measure and interpretation of the Nature.

Practically the differential calculus, complex numbers, analytic, differential and non-Euclidean geometries, the functions of a complex variable, partial differential equations and group theory emerged and were formalized.

Now we briefly see that many of these mathematical arguments are derived as an extension or a criticism of Euclidean geometry. In both cases the conclusion is a complete validation of the logical and operative cogency of Euclidean geometry.

Finally a reformulation of the finalization of Euclidean geometry has been made, so that also the "new geometries" are, today, considered in an equivalent way since they all fit with the general definition:

Geometry is the study of the invariant properties of figures or, in an equivalent way a geometric property is one shared by all congruent figures.

There are many ways to state the congruency (equivalence) between figures, here we recall one that allows us the extensions considered in this book.

Two figures F and F' are said congruent if it is possible to establish a one-to-one correspondence between their points so that the "distance" between any two points of F is the same as the distance between the corresponding points of F'. Therefore by defining **distance** in different ways we have different geometries.

In particular, the characteristic quantities of Euclidean geometry are the lengths of segments and the angles and well known examples are the theorems that state the equality between two triangles, i.e., their congruency, if three of their six characteristic elements are equal.

The steps from Euclidean geometry and this present-day definition have been realized, in three centuries, by many important scientist among which we recall: Déscartes, Gauss, Bolyai and Lobachevsky, Riemann, Beltrami, Lie, Klein.

The very interesting history is the one of the scientific development and goes over the object of this book, here we just recall the points relevant for the developments in this book.

The first step, new with respect to Euclidean geometry, has been performed by Déscartes in the 17th century. He has shown that Euclidean geometry can be based on the concept of number and reduced to analysis by the introduction of a coordinate system in the plane. Then to each point it is associated a pair of numbers (x, y), its coordinates, and to each figure F, i.e., to each set of points, there corresponds a set of number pairs, the coordinates of the points of F.

For example, the set of number pairs x, y such that $(x - a)^2 + (y - b)^2 = r^2$ corresponds to the set of points P at a fixed distance r from a given point $C \equiv (a, b)$, i.e., the circle with center C and radius r.

This step can be considered an improvement and not a criticism of Euclidean geometry. Actually for millennia it was thought that the sole aim of geometry is the exploration of the properties of three-dimensional space and there was not a shadow of doubt about the suitability of Euclidean geometry for this purpose. The criticism occurred at the beginning of 19th century with a radical change of the basic concepts, with the intuitions and formalizations of *non-Euclidean geometry* by Gauss, Bolyai and Lobachevsky. These geometries, after years of disbelief, have been corroborated by Beltrami who, by using the formalization by Gauss and Riemann of differential geometry, has given *An Euclidean interpretation of non-Euclidean geometry*.

Now we summarize the peculiar points of this route and the Gauss and Beltrami's papers.

A.1 F. Bolyai and N. I. Lobachevsky: Non-Euclidean Geometry

At the beginning of 19th century, after millennia of unquestionable acceptance, the postulates of Euclidean geometry began to be discussed, in particular the 5th one. This postulate can be set in many equivalent ways and the most accepted today is the following one: *From an external point of a straight line we can trace just one non-crossing straight line (parallel line)*.

After a lot of unsuccessful attempts for deriving this postulate from the others, Bolyai, Lobachevsky and Gauss, in an independent way, formulated a theory with the same consistency of Euclidean geometry, with a different postulate: *From an external point of a straight line we can trace an infinite number of non-crossing straight lines (parallel lines).*

This geometry has been called *non-Euclidean geometry.*

It can be noted that Euclid too retarded the use of this postulate and we can presume that he was not completely sure of its unquestionable validity. Actually the geometry was formalized for the measures of the Earth and the Euclidean geometry does not hold, as we are going to see, on the spherical surface of the Earth.

Otherwise before considering this new geometry as a rightful branch of mathematics an answer to the doubts of its coherence was necessary, i.e., while Euclidean geometry has been validated for more than 2000 years, a geometry with this new postulate could generate some contradictory results.

It's thanks to Beltrami that this doubt was eliminated. Actually he has shown that non-Euclidean geometry is equivalent with Euclid's geometry on a particular surface, therefore from the coherence of this one it follows the coherence of non-Euclidean geometry.

A.2 C. F. Gauss: Differential Geometry on Surfaces

Here we recall some concepts of differential geometry, starting from a simple example.

It is known that if we have two points on a plane, they can be connected by means of a straight line. This straight line also represents the shortest line between the given points. For this reason the length of the segment between two points is taken as their distance. In a Cartesian representation this distance is calculated by means of Pythagoras' theorem.

If instead of a plane we have an arbitrary surface we cannot go, remaining on the surface, from the first to the second point with a straight path. Otherwise between the lines connecting the two points there is, usually, a shortest one. This line is called *geodesic line.* From this definition the geodesic lines can be considered as the straight lines on a plane.

Now we may ask if it is possible to formalize this property: the answer is positive and has been given by Gauss in the fundamental work (1828): *Disquisitiones generales circa superficies curvas.*

This work originates from a practical problem: to measure the terrestrial meridian. Actually the terrestrial globe is the simplest example of a non-plane surface. As it was usual in many works of Gauss, the problem was tackled in a more general way and the obtained results are today considered one of the background of modern mathematics.

We recall some points: Gauss considers a surface in Euclidean space referred to Cartesian axes. This surface can be represented by writing the point coordinates as function of the parameters u, v:

$$x = x(u, v); \quad y = y(u, v); \quad z = z(u, v). \tag{A.1}$$

By considering two points at infinitesimal distance, their distance, from Pythagoras' theorem is given by

$$ds^2 = dx^2 + dy^2 + dz^2. \tag{A.2}$$

If we can move just on the surface, the coordinates are linked by (A.1), therefore ds^2 must be written as function of the variables (u, v) and of their differentials du, dv:

$$dx = \frac{\partial x}{\partial u} du + \frac{\partial x}{\partial v} dv; \quad dy = \frac{\partial y}{\partial u} du + \frac{\partial y}{\partial v} dv; \quad dz = \frac{\partial z}{\partial u} du + \frac{\partial z}{\partial v} dv. \tag{A.3}$$

By substituting (A.3) in (A.2), Gauss obtained the metric (or line) element

$$ds^2 = E(u, v) du^2 + 2F(u, v) du\, dv + G(u, v) dv^2. \tag{A.4}$$

The variables u, v have been called *curvilinear coordinates on the surface*.

Equation (A.2) represents a distance in a Euclidean space homogeneous in all the points, (A.3) depends on the points we consider. Therefore in a representation on a plane, as we see in Sect. A.3, it cannot be measured by a standard length. Actually it must be measured by the shortest line between the ones connecting the given points.

Gauss demonstrated that (A.4) describes completely the surface, i.e., the finite distance between two points, the area of a zone, the angle between two curves, can be determined from E, F, G and their derivatives.

From (A.4) we can also obtain the geodesic equations and it can be shown that geodesic lines have other properties in common with straight lines

1. two points determinate, in general, just one geodesic
2. from all points originate geodesics in any direction
3. a geodesic is determined by two points or one point and a direction.

Otherwise the surfaces can be considered as defined by (A.4) independently of the starting Euclidean space.

Gauss formulated the completely new concept, that a surface can be considered as a "two-dimensional space" non homogeneous in which the motions of finite figures are not in general possible.

These concepts have been generalized by Riemann to N-dimensional spaces opening another fundamental research field.

Another important idea of Gauss was *to imagine the surface as a thin voile that can be bent without expansion*. The figures on the surface, after these deformations, shall have different shapes; otherwise the finite distance between two points, the area of a figure, the angle between two curves, remain the same. These properties, that can be studied without going out of the surface, as the plane geometry without going out of the plane, constitute the *intrinsic geometry*.

From an analytic point of view these deformations are described by variables transformations $u, v \Rightarrow u', v'$.

Then, after having described the surfaces in this "intrinsic way", Gauss began to study the properties that were peculiar of each surface. With these studies he determinated the first invariant the *total curvature*, also called the *Gauss curvature*.

We recall its definition. Let us consider a point on a surface, its normal line and the sheaf of planes it determines. The lines given by the intersection of the surface with these planes, have, in the considered point, a tangent circle. The length of its radius is called *radius of curvature*. There are two curves, that result to be orthogonal, for which the corresponding radius are minimum and maximum, respectively. Let us call them r_1 and r_2: the quantity

$$K = \frac{1}{r_1 r_2} \tag{A.5}$$

is called *absolute curvature K*. Gauss finds that this quantity is the same for all the representations of the surface, i.e., it is invariant with respect to the transformation of variables u, v. Gauss called this property *Theorema Egregium*.

Now we consider an example that clarify some concepts. Let us consider an infinite circular cylinder and develop it on a plane on which it covers a band. This is an example of transformation of a surface into another one with deformation (bending without expansion).

Now we calculate the radius of curvature:

- for the plane the intersection curves with the normal planes are straight lines, for which $r_1 = r_2 \equiv \infty$ therefore $K = 0$;
- for the circular cylinder we have $r_1 = R$ (radius of the cylinder) $r_2 = \infty$ therefore once again $K = 0$.

We see that r_1 and r_2 change, while K is the same.

Also K can be obtained from the coefficients E, F, G of the metric elements and their first and second derivatives. In general its value is different for different points.

We conclude this section with an argument that will be relevant in the next one. We know that a peculiar characteristic of geometrical figures in Euclidean plane is that they can be moved in every position. A similar situation happens on a spherical surface. Are there other surfaces on which this same property holds? The reply to this question is: a general motion of a figure on a surface is possible only if the surface has constant curvature. In particular we have

- the surfaces with $K = 0$ are equivalent to a plane,
- the surfaces with $K > 0$ are equivalent to a sphere,
- the surfaces with $K < 0$ are equivalent to a surface called *pseudo-sphere* by Beltrami.

Now we see that on these last surfaces the axioms of non-Euclidean geometry are satisfied.

A.3 E. Beltrami: Essay on the Interpretation of Non-Euclidean Geometry (*Saggio di interpretazione della geometria non-euclidea*)

In a paper of 1865 Beltrami set the problem: *to seek the surfaces that can be represented into a plane so that their geodesic lines are straight lines.*

Following the Gauss approach, Beltrami formulated the problem, looking for the metric elements that satisfy the required conditions. He obtained

$$ds^2 = R^2\frac{(a^2 + v^2)du^2 - 2u\,v\,du\,dv + (a^2 + u^2)dv^2}{(a^2 + u^2 + v^2)^2} \tag{A.6}$$

$$ds^2 = R^2\frac{(a^2 - v^2)du^2 + 2u\,v\,du\,dv + (a^2 - u^2)dv^2}{(a^2 - u^2 - v^2)^2} \tag{A.7}$$

with R, a arbitrary constants. By means of Beltrami words, we summarize this result:

◁ For recognizing the nature of these surfaces we calculate the total (Gauss) curvature, finding for (A.6) $K = 1/R^2$ and for (A.7) $K = -1/R^2$, therefore we can conclude: The only surfaces that can be bijectively mapped into a plane so that their geodesic lines become straight lines are the ones with constant curvature: positive, negative or zero ▷.

The last result corresponds to a plane, therefore it was well known. Also the first one was known. It represents a spherical surface, which geodesic lines are given by the maximum circles. These circles are the intersection of the surface with planes passing through the center. Therefore by projecting the points of the surface from the center into a tangent plane, the maximum circles are mapped into straight lines.

Anyway the used general method and the extension to negative constant curvature surfaces (A.7), are new.

By means of this theorem the paper is complete, but Beltrami looked for a geometrical interpretation of his results and (1868) the "*Saggio*" originated. The motivations for this work are reported in its introduction:

◁ In these last times the mathematical people begins to consider new concepts that, if they are true, will change all the classical geometry. These efforts for a radical change of the principles are common in the history of the knowledge. Moreover they are today consequent on the right critical examinations of scientific investigations.

If these efforts originate from conscientious researches and genuine belief the duty of science people is to discuss them impartially equally far from enthusiasm and contempt. For these reason we have investigated about the results obtained from Lobachevsky theory, and we tried to find a real interpretation before to admit that a new order of concepts is necessary. ▷

After this introduction, Beltrami inquires into the basic meaning of Euclidean geometry and extends it: ◁ The fundamental criterion of demonstration of elementary geometry is the *possibility of superimposing equal figures*. This criterion not only applies to plane, but also to the surfaces on which equal figures may exist in different positions, as an example this property holds on the spherical surfaces unconditionally... The straight line is a fundamental element in figures and constructions of elementary geometry. A particular characteristic of this element is that it is completely determined by just two of its points... This particular feature not only characterizes straight lines on the plane, but also is peculiar to geodesic lines, that represent, for the surfaces the curves of minimal lengths, i.e., the generalizations of the straight lines. ▷

In particular ◁ an analogous property of straight lines completely applies to spherical surfaces, that is: if two spheres with the same radius are given and a geodesic line exists on each surface, superimposing one surface on the other, in such a way that the geodesics are superimposed in two points, they are superimposed along all their extension. The analogies between spherical and plane geometries are based on the aforesaid property.

These considerations have been the starting reasons for the present researches. Actually the same geometrical considerations, which are evident for the sphere, may be extended to another surface in Euclidean space which can be called **pseudo-sphere**. A surface with this property is obtained by rotating the "constant tangent curve" also called "tractrix" Fig. A.1 (I). ▷ Let us see some peculiarities of this geometry starting from the metric element (A.7). We have

- ◁ the values for the variables u, v, are limited from the relation $u^2 + v^2 \leq a^2$, i.e., if we consider u, v as Cartesian coordinates on a plane, the tractrix is represented inside a circle that is called *limiting circle*:

$$u^2 + v^2 = a^2 \tag{A.8}$$

  In this representation the geodesics of the surface are the chords of the limiting circle Fig. A.1 (II).

- By considering the distance r of a point from the origin, its value increases from 0 for $r = 0$, and infinite for $r = a$, i.e., for the values of u, v satisfying (A.8). Therefore the circumference (A.8), represents the points of the surface at infinite distance. ▷

Therefore by considering a chord and an external point, from this point we can draw infinite "straight lines" that cross the given chord and infinite that do not cross it and two that cross it on the limiting circle, i.e., in the points at infinite distance, and can be called *parallel straight lines*, see Fig. A.1 (II).

Since these chords represent the geodesic of the pseudo-sphere we say that from an external point of a geodesic can be drawn an infinite number of geodesics crossing it and an infinite number of geodesics that do not cross it. We observe that while for the sphere there is a geometrical mapping (projection from the center into a tangent plane) that allows to transform the geodesics into straight lines, for negative constant curvature surfaces the representation of Beltrami is obtained by an analytical transformation of variables.

After a detailed comparison of the theorems of Lobachevsky's non-Euclidean geometry with the equivalent ones on geodesic lines, Beltrami concludes

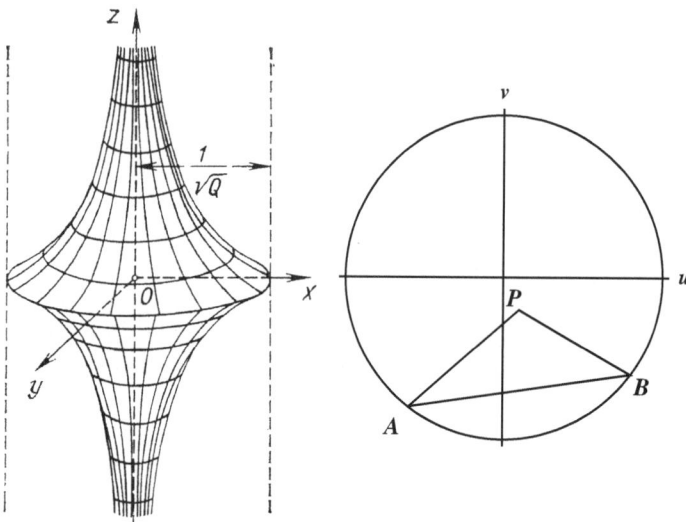

Fig. A.1 The pseudo-sphere (I) (on the left) and representation on a plane of non-Euclidean geometry (II) (on the right). In a Euclidean plane the segments and straight lines have many relevant properties. In particular, by means of them we construct the geometrical figures and we measure the distances between two points. The concept of straight line loses its meaning if we consider a general surface. Actually for calculating, on these surfaces, the distance between two points, the geodesic lines are introduced. These lines also hold other properties peculiar of straight lines. Beltrami demonstrated that on a particular surface, called pseudo-sphere (I), the geodesic lines satisfy to postulates of non-Euclidean geometry of Lobachevsky. By means of this "euclidean representation" he demonstrated that an *abstract geometry* holds the same cogency of Euclidean geometry. Actually, following Gauss, all surfaces can be represented into a plane u, v (or a part), where u, v are the parameters (curvilinear coordinates) that identify the points of the surfaces in the space (A.1). On this plane (II) the distance between points is calculated by means of the geodesic lines of the surface, obtained from the metric element introduced by Gauss (A.4). All bijective transformations of the variables $u, v \leftrightarrow u', v'$ change the geodesic equations. Beltrami found the transformation by which the geodesic lines of a negative constant curvature $(-K = Q)$ surface, are represented by the chords of a circle. If we consider the chord $\overline{AB}$ and an external point P, from this point we can draw infinite "straight lines" (chords) that cross $\overline{AB}$, infinite that do not cross it and two, $\overline{PA}$ and $\overline{PB}$, that cross it on the limiting circle. Since these points represent the points at infinite distance, they are considered as parallel straight lines. The straight lines (chords) and the triangles satisfy the theorems of Lobachevsky non-Euclidean geometry

- ◁ The theorems of non-Euclidean geometry (of Lobachevsky) for the plane figures realized with straight lines, hold for the same figures realized with the geodesic lines on the pseudo-sphere.
- A complete correspondence between non-Euclidean planimetry with the geometry on the pseudo-sphere exists. ▷

Then ◁ results that seemed to be inconsistent with the hypothesis of a plane become consistent with a surface of the above-mentioned kind and, in this way, they can get a simple and satisfactory explanation. ▷

Therefore by substituting the Euclide's fifth postulate, we have

1. *From an external point of a straight line (geodesic) we cannot draw a parallel line (geodesic).*
   The results obtained are the same of the Euclidean geometry on a spherical surface, for which all the geodesics lines (maximum circles) cross in diametrically opposite points.
2. *From an external point of a straight line (geodesic) we can draw infinite parallel lines (geodesics).*
   The results obtained are the same of the Euclidean geometry on a pseudo-spherical surface Fig. A.1.

The non-Euclidean plane geometry acquires the same coherence and relevance of Euclidean geometry. Beltrami concludes:

◁ With the present work we have offered the development of a case in which abstract geometry finds a representation in the concrete (Euclidean) geometry, but we do not want to omit to state that the *validity of this new order of concepts* is not subordinate to the existence (or not) of such a correspondence (as it surely happens for more than two dimensions). ▷

The relevance of this work goes over the already remarkable mathematical meaning, actually it represents a milestone for the scientific development. Now we recall this very important epistemological aspect: the formulation of Euclidean geometry starts from postulates which are in agreement with the experience and all the following statements are deduced from these postulates (axiomatic-deductive method). Lobachevsky uses the same axiomatic-deductive method, but he starts from "abstract" postulates, which do not arise from experience. The work of Beltrami demonstrates that the theorems of Lobachevsky are valid on a surface of Euclidean space, i.e., *also starting from arbitrary axioms, we can obtain results in agreement with our "Euclidean" experience.*

These concepts can be generalized and applied to scientific research: arbitrary hypotheses acquire validity if the consequent results are in agreement with experience. The most important discoveries of modern physics (Maxwell equations, special and general relativity, quantum theory, Dirac equation) arose from this "conceptual broad-mindedness".

We conclude this section recalling the last formalization of non-Euclidean geometry.

By means of a particular variable transformation $u, v \rightarrow u', v'$ Felix Klein showed that Beltrami's geodesic lines of Fig. A.1 (II) are mapped into arcs of circumferences orthogonal to the circle limit.[1]

---

[1] In four woodcuts, called "Circle Limit I–IV", Escher represented these geodesics and some figures. In Circle Limit III, following the scheme of the mathematician D. Coxeter [1], he perfectly shows as the figures (fishes) with the same dimension appear, in this Euclidean representation, deformed and reduced as they approach the circle limit. In fact drawing closer to circle limit the distances grow to "infinite values" and the figures look as "infinitesimal".

Starting from Klein formulation and using a transformation with a function of a complex variable, Henri Poincaré extended into the infinite zone of a half plane the non-Euclidean geometry limited by Klein's circle. This extension to an infinite zone (not limited by Beltrami or Klein circle) eliminated another doubt about the cogency of this geometry.

The final step for the present-day definition of geometry is due to Klein. Before discussing this point we recall another very important idea of the modern mathematics and physics: the concept of group and, in particular *the transformation groups* introduced by Sophus Lie.

A.4 S. Lie: The Continuous Transformation Groups

The groups of transformations were introduced and formalized by S. Lie in the second half of the 19th century. The concept of group is, today, well known; here we only recall the definition of transformations or Lie groups that is used in this book.

Let us consider N variables x^m and N variables y^i that are functions of x^m and of K parameters a^l. Now, let us consider a second transformation from y^i to the variables z^n, given by the same functions but with other values of the K parameters b^l. We say that these transformations are a group if, by considering them one after the other (product of two transformations), we have the same functional dependence between z and x with parameters, whose values depend just on the parameters of the previous transformations. To express these concepts in formulas, we indicate by

$$y^i = f^i(x^1, \ldots, x^N, a^1, \ldots, a^K) \equiv f^i(x, a) \quad \text{with } i = 1, \ldots, N$$

the first transformation and by $z^n = f^n(y, b)$ a second transformation with other parameters. The relation between z and x (composite transformation) is $z^n = f^n[f(x, a), b]$. If these last transformations can be written as $z^n = f^n(x, c)$ with $c = g(a, b)$, they represent a group. The number of parameters (K) is the order of the group. If the parameters can assume continuous values, these groups are called continuous groups.

A.5 F. Klein: Linear Transformations and Geometries (Erlanger Programm)

In an inaugural lecture today known as "Erlanger Programm", Felix Klein in 1872 associated the transformation groups with geometries. The guideline is the following: the notion of equivalence is associated with all the geometries (elementary, projective, etc.). It can be shown that the transformations that give rise to equivalent forms have the properties of groups. Then, the geometries are the theories related to the invariants of the corresponding groups. From these general considerations Klein points out the relevance of linear transformations as related to geometries which have the property of describing homogeneous spaces, i.e., spaces in which the properties are the same in all their points.

An arbitrary linear transformation can be written as

$$y^\gamma = \sum_{\beta=1}^{N} c^\gamma_\beta x^\beta; \quad c^\gamma_\beta = \text{constants} \quad \text{and} \quad \|c^\gamma_\beta\| \neq 0;$$

therefore it depends on N^2 parameters. By identifying y^γ and x^β as vector components, we can write, following the notation of linear algebra,

$$\begin{pmatrix} y^1 \\ \vdots \\ y^N \end{pmatrix} = \begin{pmatrix} c^1_1 & \cdots & c^1_N \\ \vdots & \ddots & \vdots \\ c^N_1 & \cdots & c^N_N \end{pmatrix} \begin{pmatrix} x^1 \\ \vdots \\ x^N \end{pmatrix}. \tag{A.9}$$

These transformations are known as **homographies** and are generally non-commutative. From a geometrical point of view, in a Cartesian representation, they correspond to take a new reference frame with arbitrary rotation and change of length measures of each axis.

The geometry that considers equivalent the geometric figures with respect to these transformations is called **affine geometry**, and the corresponding group is called **affine group**. An invariant quantity for such a group is the distance between two points, expressed by a positive definite quadratic form (the distances are obtained by means of a generalization to N-dimensional spaces, of Carnot's theorem).

Now let us consider the familiar **Euclidean geometry** for which the basic properties of the figures are the length of segments and angles. These elements are unchanged by rotations and translations of the figures. Therefore these are the allowed transformations in Euclidean geometry.

This geometry can also be considered as a subgroup of the affine group for which the distance is given by Pythagoras' theorem. For the Euclidean group the constants c^γ_β of (A.9) must satisfy $N(N+1)/2$ conditions, then the remaining parameters are $N^2 - N(N+1)/2 = N(N-1)/2$.

From a geometrical point of view these transformations, in a Cartesian representation, are given by the rotations of the frame of reference.

If we consider also the **translations group** (the translation of coordinate axes origin), we must add N parameters and we get a group with $N(N+1)/2$ parameters, which corresponds to the allowed `motions` of geometrical figures in a N-dimensional Euclidean space.

Conclusions

The ultimate result of this development of mathematics has been that the spreading out in many branches and the uncertainty at the beginning of 19th century are disappeared at the end of the century so that Hilbert could summarize: Mathematics is an organism who keeps its vital energy from the indissoluble ties between its various parts.

We have shown in this book that the physics too is present in this context allowing the introduction of the "geometry of space-time" and its complete formalization by means of the generalization of complex numbers.

Reference

1. S. Roberts, *King of Infinite Space: Donald Coxeter, the Man Who Saved Geometry* (Walker and Company, 2006)

Index